SpringerBriefs in Research and Innovation Governance

Editors-in-Chief

Doris Schroeder, *Centre for Professional Ethics, University of Central Lancashire, Preston, UK*
Konstantinos Iatridis, *School of Management, University of Bath, Bath, UK*

SpringerBriefs in Research and Innovation Governance present concise summaries of cutting-edge research and practical applications across a wide spectrum of governance activities that are shaped and informed by, and in turn impact research and innovation, with fast turnaround time to publication. Featuring compact volumes of 50 to 125 pages, the series covers a range of content from professional to academic. Monographs of new material are considered for the SpringerBriefs in Research and Innovation Governance series. Typical topics might include: a timely report of state-of-the-art analytical techniques, a bridge between new research results, as published in journal articles and a contextual literature review, a snapshot of a hot or emerging topic, an in-depth case study or technical example, a presentation of core concepts that students and practitioners must understand in order to make independent contributions, best practices or protocols to be followed, a series of short case studies/debates highlighting a specific angle. SpringerBriefs in Research and Innovation Governance allow authors to present their ideas and readers to absorb them with minimal time investment. Both solicited and unsolicited manuscripts are considered for publication.

Kate Chatfield · Michelle Singh
Editors

Research Ethics and Integrity During Pandemics

Developing the PREPARED Code

 Springer

Editors
Kate Chatfield
Centre for Professional Ethics
University of Central Lancashire
Preston, UK

Michelle Singh
European and Developing Countries Clinical
Trials Partnership (EDCTP) Association
Cape Town, South Africa

ISSN 2452-0519　　　　　　　ISSN 2452-0527　(electronic)
SpringerBriefs in Research and Innovation Governance
ISBN 978-3-031-91323-5　　　ISBN 978-3-031-91324-2　(eBook)
https://doi.org/10.1007/978-3-031-91324-2

This work was supported by University of Central Lancashire.

© The Editor(s) (if applicable) and The Author(s) 2025. This book is an open access publication.

Open Access This book is licensed under the terms of the Creative Commons Attribution 4.0 International License (http://creativecommons.org/licenses/by/4.0/), which permits use, sharing, adaptation, distribution and reproduction in any medium or format, as long as you give appropriate credit to the original author(s) and the source, provide a link to the Creative Commons license and indicate if changes were made.
The images or other third party material in this book are included in the book's Creative Commons license, unless indicated otherwise in a credit line to the material. If material is not included in the book's Creative Commons license and your intended use is not permitted by statutory regulation or exceeds the permitted use, you will need to obtain permission directly from the copyright holder.
The use of general descriptive names, registered names, trademarks, service marks, etc. in this publication does not imply, even in the absence of a specific statement, that such names are exempt from the relevant protective laws and regulations and therefore free for general use.
The publisher, the authors and the editors are safe to assume that the advice and information in this book are believed to be true and accurate at the date of publication. Neither the publisher nor the authors or the editors give a warranty, expressed or implied, with respect to the material contained herein or for any errors or omissions that may have been made. The publisher remains neutral with regard to jurisdictional claims in published maps and institutional affiliations.

This Springer imprint is published by the registered company Springer Nature Switzerland AG
The registered company address is: Gewerbestrasse 11, 6330 Cham, Switzerland

If disposing of this product, please recycle the paper.

In memory of
Dr Vasantha Muthuswamy (1948–2023)

Foreword

As a leader of a major research funding institution with a global dimension, my focus is always on how to encourage excellent collaborative research that benefits humanity. There is no doubt for me that excellent research must also be ethical research, irrespective of location. Unethical research can create harm, for instance to research participants, but it can also destabilise the scientific community by supporting unhealthy research environments, and it can catalyse public mistrust.

Ethics codes play an important role in ensuring that research can be conducted with high scientific, regulatory and ethical standards. For instance, at the European & Developing Countries Clinical Trials Partnership (EDCTP), we successfully use the TRUST Code (TRUST 2018) to stop helicopter research and ethics dumping, that is, the export of unethical practices from higher- to lower-income settings. We currently have a budget of €1.86 billion to fund global health projects through EDCTP3,[1] and *equitable* research partnerships between our 30 African and 15 European country partners are crucial to our success.

As an African and European research partnership on infectious diseases, EDCTP also deals prominently with research to prevent disease outbreaks, epidemics and pandemics. For instance, in 2024 we launched an emergency research response call on mpox that is supporting cross-border research spanning nine African countries. Moreover, early in 2025 we announced that we are funding additional research projects to combat the mpox epidemic.[2]

I am very grateful that we now have a focused ethics code that can be applied during a pandemic but is also likely to be useful during epidemics and public health emergencies of international concern.

What is particularly noteworthy is that the PREPARED Code includes all the benefits of the TRUST Code: it is short, jargon-free and values-driven, applies to all research disciplines and was developed in a highly inclusive manner, but now adds a new achievement, namely the combination of research ethics and research integrity guidance in one code. For me as the Executive Director of a major research funding institution, it is important that the PREPARED Code has abandoned the division or silo building between research ethics and research integrity. I believe that this will lead to a healthier research environment, governed by fairness, respect, care and honesty.

This book provides insights into the development of an ethics code which are highly illuminating and might even encourage others to try the same. While the PREPARED Code is only three pages long, this book reads like a complex research adventure, describing all the intricate steps necessary to develop a credible ethics code. I congratulate the editors, Dr Kate Chatfield and Dr Michelle Singh, as well as all authors, on the book and I congratulate the lead author of the TRUST and the PREPARED Codes, Prof. Doris

---

[1] https://www.global-health-edctp3.europa.eu/.

[2] https://www.edctp.org/news/global-health-edctp3-funds-additional-research-projects-to-combat-mpox/.

Schroeder, on adding to the family of ethics codes that try to reach everybody through globally understood moral values.

The world would be a better place if more human activities were governed by fairness, respect, care and honesty.

Michael Makanga

Reference

TRUST (2018) The TRUST code: a global code of conduct for equitable research partnerships. https://doi.org/10.48508/GCC/2018.05. Accessed 5 February 2025.

Dr Michael Makanga is a clinician-scientist with nearly thirty years of health and clinical research work experience in African and European institutions and Executive Director of the Global Health EDCTP3 Joint Undertaking. He has a medical degree from Makerere University, a master's from Liverpool University and a PhD in Pharmacology and Therapeutics from the Liverpool School of Tropical Medicine and is a fellow of the Royal College of Physicians in Edinburgh. He has served in various scientific and policy advisory boards involved in developing medical products and associated technologies for infectious diseases, including the World Bank, international product development organisations, philanthropies and pharmaceutical companies. He is an adviser to the PREPARED project and a co-author of the TRUST and the PREPARED Codes.

Acknowledgements

This book is an output of the PREPARED project funded by the European Commission, UK Research and Innovation (UKRI) and the Swiss State Secretariat for Education, Research and Innovation (SERI) (grant number 10048353).[3] The project team received not only funding, but also considerable support, from the funding organisations, in particular from Dorian Karatzas, Roberta Monachello, Mihalis Kritikos and Lisa Diependale at the European Commission, and Raphael Misteli and Brita Bamert at SERI. We are extremely grateful for this funding and support, as it helped us develop tools to ensure that accelerated research does not violate research ethics and integrity values. One of these tools is the **PREPARED Code: A Global Code of Conduct for Research during Pandemics.**

We thank everybody who worked for more than two years, from September 2022 to December 2024, to develop the code.

The PREPARED team comprised almost 60 people from 16 organisations around the world. We thank each of them for always bringing their best to the mission.

Our profound thanks to the advisers who were involved with the PREPARED Code: Dr Raffaella Ravinetto, Dr Shunzo Majima, Dr Johannes Rath, Prof. Fatima Alvarez-Castillo, Dr Fritz Schmuhl, Dr Sally Dalton-Brown, Dr Jantina de Vries, Emma Law, Dr Michael Makanga, Dr João Monteiro and Dr Saleh Aljadeeah.

We thank members of all eight PREPARED advisory platforms (funders, policymakers, industry, researchers, citizens, nongovernmental organisations, publishers and governance) for their support during the development and refinement of the code.

We are indebted to the following individuals for providing elegant translations of the PREPARED Code: Dr Carlos Almeida Pereira, Lusophone Platform for Clinical Research and Biomedical Innovation (PLICIB) (Portuguese), Prof. Doris Schroeder (German), Dr Saleh Aljadeeah (Arabic), Linzi Zhu (Chinese), Dr Young Su (Korean), Prof. Stéphanie Laulhé-Shaelou (French), Clàudia Pallisé Perelló (Spanish), Dr Andreas Marcou and Dr Katerina Kalaitzaki (Greek), Dr Giulia Inguaggiato (Italian), Dr Kalle Videnoja (Finnish), Rhoda Kabuti (Swahili), and Vilma Lukaševičienė and Jūratė Lekstutienė (Lithuanian). A big thank you also to the quality checkers: Dr Lisa Tombornini (German), Prof. Wei Zhu (Chinese), Prof. Ock-Joo Kim (Korean), Dr Francois Bompart (French), Dr Dafna Feinholz (Spanish), Anni Sairio (Finnish), Joyce Odhiambo and Dr Joshua Kimani (Swahili), Dr Nearchos Paspallis (Greek), and Prof. Eugenijus Gefenas and Vygintas Aliukonis (Lithuanian).

Completing the code in a little over two years and spending the past five months writing this book meant we were running two overlapping races. We would like to thank

---

[3] The views and opinions expressed are, however, those of the authors only and do not necessarily reflect those of the European Union, the European Research Executive Agency, UKRI or SERI. Neither the European Union, the granting authority, UKRI nor SERI can be held responsible for them.

the other 23 contributing authors for keeping to their deadlines and producing high-quality chapters. You were an enthusiastic team, and it was a pleasure to work with each of you: Joyce Adhiambo Odhiambo, Pamela Andanda, Orla Drummond, Natalie Evans, Dafna Feinholz, Eugenijus Gefenas, Giulia Inguaggiato, Ock-Joo Kim, Joshua Kimani, Nandini Kumar, Emma Law, Klaus Leisinger, Vilma Lukaševičienė, Langelihle Mlotshwa, Thomas Nyirenda, Hazel Partington, Nearchos Paspallis, Clàudia Pallisé Perelló, David Robinson, Doris Schroeder, Carly Seedall, Kalle Videnoja and Wei Zhu.

Our very efficient and helpful Senior Editor at Springer, Doris Bleier, made the process—from discussing the book proposal, through responding to reviewer comments, to contracting and then finally delivering the book—seem easy. Our Production Editor, Saranya Kalidoss, also helped smooth the process significantly, for which we are very grateful.

This was our first collaboration with Paul Wise, the copy editor from South Africa, and all the glorious reports we had heard about him are certainly true. His editing skills are unsurpassed. Thank you, Paul.

Thank you to our talented artist, David Robinson (https://www.daverob.co.uk/), for his creative and inspiring graphics, which make this book a more engaging read.

A special thank you to Dr Michael Makanga, the Executive Director of Global Health EDCTP3, who wrote an apt and beautiful Foreword despite being one of the busiest people we know.

Our immense gratitude to the lead author of the PREPARED Code, Prof. Doris Schroeder, for her insight, creativity and support throughout this endeavour.

This book is dedicated to Dr Vasantha Muthuswamy (1948–2023), who started working on the code as part of the PREPARED team, but then passed away much too early, and is greatly missed.

We hope the book gives an insight into the development of risk-based ethics codes, as well as enough information on the PREPARED Code to ensure that we are better prepared for the next pandemic.

March 2025   Kate Chatfield
                                                                                                                                Michelle Singh

Contents

Research Ethics and Integrity During Pandemics: Not Unique, but Vastly Magnified Challenges 1
   Kate Chatfield and Michelle Singh

The PREPARED Code: A Global Code of Conduct for Research During Pandemics 8
   Doris Schroeder and David Robinson

Ensuring Effectiveness and Credibility: The Conceptual Foundation of the PREPARED Code 16
   Kate Chatfield, Doris Schroeder, Eugenijus Gefenas, Vilma Lukaševičienė, Kalle Videnoja, Emma Law, Joyce Adhiambo Odhiambo, and Joshua Kimani

Research Ethics and Integrity Challenges During Pandemics: The Research Foundation of the PREPARED Code 34
   Pamela Andanda, Langelihle Mlotshwa, Orla Drummond, Vilma Lukaševičienė, Giulia Inguaggiato, and Klaus Leisinger

From Real-World Challenges to a Global Code: How the PREPARED Code Was Built 53
   Natalie Evans, Hazel Partington, Doris Schroeder, Kate Chatfield, Clàudia Pallisé Perelló, Nandini Kumar, Ock-Joo Kim, Wei Zhu, and Dafna Feinholz

Implementation Support for the PREPARED Code 76
   Carly Seedall, Doris Schroeder, Nearchos Paspallis, Thomas Nyirenda, and Michelle Singh

Learning from the PREPARED Experience: Recommendations for Enhancing the Effectiveness and Credibility of New Ethics Codes 94
   Kate Chatfield and Michelle Singh

Author Index 103

Editors and Contributors

About the Editors

Kate Chatfield is an empirical ethicist at the Centre for Professional Ethics, University of Central Lancashire (UCLan) UK. She has a background in philosophy and bioethics and extensive experience in social science research. As an empirical ethicist, her research has been focused upon research ethics, research integrity and global justice.

Michelle Singh is a biomedical scientist at the European and Developing Countries Clinical Trials Partnership (EDCTP) Association, South Africa. She holds a doctorate in obstetrics and gynaecology, a master's in biomedical science and a master's in global bioethics. She is a co-author of the TRUST Code and partner in the PREPARED research ethics and integrity consortium.

List of Contributors

Pamela Andanda School of Law, University of the Witwatersrand, Johannesburg, South Africa

Kate Chatfield Centre for Professional Ethics, University of Central Lancashire, Preston, UK

Orla Drummond Trilateral Research, Belview Port, Ireland

Natalie Evans Department of Ethics, Law and Humanities, Amsterdam UMC, Vrije Universiteit Amsterdam, Amsterdam, The Netherlands

Dafna Feinholz United Nations Educational, Scientific and Cultural Organization, Paris, France

Eugenijus Gefenas Faculty of Medicine, Vilnius University, Vilnius, Lithuania

Giulia Inguaggiato Department of Ethics, Law and Humanities, Amsterdam Public Health Institute, Amsterdam UMC, Vrije Universiteit Amsterdam, Amsterdam, The Netherlands

Ock-Joo Kim Seoul National University, Seoul, South Korea

Joshua Kimani Partners for Health and Development in Africa, Nairobi, Kenya

Nandini Kumar Forum for Ethics Review Committees in India, New Delhi, India

Emma Law PROTAS, London, UK

Klaus Leisinger Foundation Global Values Alliance, Basel, Switzerland

Vilma Lukaševičienė The Division of Medical History and Ethics in the Medical Faculty of Vilnius University, Vilnius University, Vilnius, Lithuania

Langelihle Mlotshwa School of Law, University of the Witwatersrand, Johannesburg, South Africa

Thomas Nyirenda European and Developing Countries Clinical Trials Partnership Association, Cape Town, South Africa

Joyce Adhiambo Odhiambo Partners for Health and Development in Africa, Nairobi, Kenya

Hazel Partington Centre for Professional Ethics, University of Central Lancashire, Preston, UK

Nearchos Paspallis Interdisciplinary Centre for Law, Alternative and Innovative Methods, Larnaca, Cyprus; UCLan Cyprus, Larnaca, Cyprus

Clàudia Pallisé Perelló Department of Ethics, Law and Humanities, Amsterdam UMC, Vrije Universiteit Amsterdam, Amsterdam, The Netherlands

David Robinson DaveRob Design and Illustration, Preston, UK

Doris Schroeder School of Law, UCLan Cyprus; Centre for Professional Ethics, University of Central Lancashire, Preston, UK

Carly Seedall European Network of Research Ethics Committees, Bonn, Germany

Michelle Singh European and Developing Countries Clinical Trials Partnership Association, Cape Town, South Africa

Kalle Videnoja Finnish National Board On Research Integrity (TENK), Helsinki, Finland

Wei Zhu Fudan University, Shanghai, China

Abbreviations

AVAREF	African Vaccine Regulatory Forum
EDCTP	European & Developing Countries Clinical Trials Partnership
ENRIO	European Network of Research Integrity Offices
EUREC	European Network of Research Ethics Committees
GloPID-R	Global Research Collaboration for Infectious Disease Preparedness
REC	Research Ethics Committee
TENK	Tutkimuseettinen neuvottelukunta (Finnish National Board on Research Integrity)
UKCDR	UK Collaborative on Development Research
UKRIO	UK Research Integrity Office
WHA	World Health Assembly
WHO	World Health Organization
WHO AFRO	WHO Regional Office for Africa
WMA	World Medical Association (WMA)

Research Ethics and Integrity During Pandemics: Not Unique, but Vastly Magnified Challenges

Kate Chatfield[1] (✉) and Michelle Singh[2]

[1] Centre for Professional Ethics, UCLan, Preston, UK
kchatfield@uclan.ac.uk

[2] European and Developing Countries Clinical Trials Partnership Association, Cape Town, South Africa

Abstract. This chapter sets the scene for the development of the PREPARED Code: A Global Code of Conduct for Research During Pandemics. Recalling the time when successive waves of the COVID-19 pandemic led to the deaths of millions and put health systems under enormous pressure, we explain how the pandemic created a demand for rapidly available, trusted scientific advice. Fast reaction systems, including accelerated research, faced significant ethics and integrity challenges. While most such challenges encountered during the COVID-19 pandemic were not unique, researchers and research ethics committees were ill-equipped to cope with their extent and scale. This chapter explains the purpose of the PREPARED Code against that backdrop, including what sets this code apart from many other research ethics codes.

Keywords: COVID-19 pandemic · research ethics · research integrity · ethics codes

1 The Purpose of the PREPARED Code

The COVID-19 pandemic presented the most challenging global health crisis in living memory (WHO 2022), triggering an urgent need for rapid research and innovation to address the far-reaching healthcare, social, cultural and economic consequences. Yet, amid the race to develop vaccines, treatments and public health interventions, a host of ethical dilemmas emerged (Barroga and Matanguihan 2020), exposing a significant gap in the existing frameworks governing research ethics and research integrity.

The chaotic rush to find solutions during the pandemic highlighted the critical importance of a robust ethical framework to guide research during global emergencies (Saxena et al. 2021). The need for an operational code that safeguards ethical values while supporting a swift and effective research response was crystal clear.

The purpose of this book is to explore the development of a pioneering ethics code designed specifically to support research ethics and integrity during pandemics. At the heart of the development was the premise that while research is essential during global health crises, it must be conducted in accordance with the highest ethical standards (Solbakk et al. 2021).

This book is an edited collection whose authors were all members of the PREPARED project team[1] that developed the PREPARED Code from September 2022 to December 2024. In the forthcoming chapters, the authors walk readers through the meticulous development of the PREPARED Code, summarising the key steps and findings.

2 An Ever-Changing Research Landscape

To begin, we invite you to cast your mind back to the early days of COVID-19, when successive waves of the disease led to the deaths of millions and put global health systems under enormous pressure (Independent Panel 2021). The pandemic created a demand for trusted scientific guidance that was unparalleled in its urgency (WHO 2022). People were desperate for effective treatments, preventative measures and public health interventions to counter the emerging and potentially devastating impacts of the pandemic. Researchers across all disciplines faced a unique combination of urgency, uncertainty and logistical hurdles. For those in health-related fields, most of whom had little or no prior knowledge of coronaviruses or epidemics, the race was on.

Over a period of just a few months, the research landscape altered dramatically, as the consequences of limited face-to-face contact were felt (Maison et al. 2021). Many research institutions and universities closed or significantly restricted on-site academic activities (Omary et al. 2020); almost all laboratory-based research, research with human participants and field research was stopped or suspended. New restrictions affected most research fields, including clinical trials, with most trials postponed or delayed (Shawrav 2022). As research projects faced delays, modifications or suspensions due to pandemic-related restrictions (Bratan et al. 2021), ongoing, non-COVID-19 studies experienced interruptions for unspecified periods. Across all research types, participants encountered changes to study methods (like switching to online communication). For some, such as elderly participants with cognitive impairments, continued participation was fraught with difficulties (Sharma et al. 2022).

It is clear from data available on ClinicalTrials.gov, a publicly accessible database of privately and publicly funded clinical trials conducted globally, that non-COVID-19 research, particularly in healthcare, was deprioritised in favour of pandemic-related studies. For instance, from January to May 2020, there was a marked decrease in the start of non-coronavirus trials, dropping from 2,616 trials in January to fewer than 1,500 trials in May (Xue et al. 2020). And for those that were ongoing, the number that were stopped averaged 1,147 trials per month (Gaudino et al. 2020). Meanwhile, the start of new COVID-19 related trials soared from 30 new trials in January 2020 to 784 new trials in April 2020 (Xue et al. 2020).

Research staff and resources were "purposely and purposefully" prioritised to COVID-19 activities above all else (Harper et al. 2020), as funding bodies and governments redirected resources towards COVID-19 research, impacting the availability

---

[1] https://prepared-project.eu/our-team/

of support for other research studies. Individual institutions also implemented new policies to address the challenges posed by the pandemic, affecting research operations and priorities (Radecki and Schonfeld 2020).

Against this backdrop, it is perhaps unsurprising that more than 50% of surveyed researchers reported poor levels of wellbeing and mental health during the COVID-19 pandemic as work changes and additional demands had a negative impact on motivation and general wellbeing (Heo et al. 2022). Furthermore, many studies reported that the burdens on some researchers, such as junior researchers and women, were greater than on others; for women, this was largely because the onus of domestic responsibilities and childcare tended to fall more heavily on them (Doyle et al. 2021).

The urgency of the crisis and the pressures upon research teams also affected the trustworthiness of research, compromising the quality, transparency and ethical standards of many research studies (Dinis-Oliveira 2020).

As researchers faced pressures to produce and publish results rapidly during the early stages of the COVID-19 pandemic, most journals in biomedicine, health and social care experienced a significant increase in the number of manuscript submissions. For example, the *Journal of the American Medical Association* reported that more than 11,000 manuscripts were submitted between 1 January and 1 June 2020, compared with approximately 4,000 during the same period in 2019, attributing virtually the entire increase to COVID-19-related manuscripts (Bauchner et al. 2020). The number of resultant publications also increased rapidly; in May 2020, *The Economist* reported that since January 2020 the number of COVID-19-related scientific publications had been doubling every 14 days, reaching 1,363 by early May (Economist 2020).

With such a high demand for publication, only a small percentage of submissions could be published in respected peer-reviewed journals, which led to a surge in preprints (studies published before peer review) (Fraser et al. 2021). While this undoubtedly facilitated rapid access to data, the fact that the preprints were not peer-reviewed allowed conclusions lacking scientific support to gain traction (Brierley 2021). Additionally, at a time when reliable evidence was desperately needed, the quality of most COVID-19 clinical studies was poor – for instance, the studies had small sample sizes or lacked rigorous methodologies – and there was a significant amount of waste in clinical research (Law and Smith 2024).

Crises can also lead to researchers cutting corners in research ethics. For instance, during the 2014 Ebola crisis in West Africa, overseas researchers carried out research among Ebola survivors without research ethics approval. This was discovered when they tried to obtain approval retrospectively in order to publish their results (Tegli 2018).

3 The Research Ethics and Research Integrity Response

For almost a century, research ethics has driven efforts to make science more ethical and to stop exploitation of and harm to research participants (Resnik 2018). For almost half that time, efforts in pursuit of research integrity have tried to achieve truthful science without fabrication, falsification or plagiarism (Zhaksylyk et al. 2023). Over this time, ethics guidelines have proliferated; the International Compilation of Human Research Standards, published by the US Department of Health and Human Services, lists over

1,000 laws, regulations, and guidelines governing human participants in research across 131 countries and numerous international organisations (HHS 2024). Yet the unprecedented scale and nature of the COVID-19 pandemic caught the world unprepared. While most of the specific challenges of research ethics and research integrity were not unique to the pandemic, researchers and research ethics committees were ill-equipped to cope with their extent and scale.

Research ethics committees found themselves with increased workloads; the urgency of COVID-19 research required them to expedite approvals for studies related to treatments, vaccines and public health interventions (Kebenei et al. 2024). Some committees found themselves confronting emerging ethical debates not previously encountered, for instance regarding the permissibility of human challenge studies, in which healthy volunteers would be deliberately infected with the infective agent to study the impacts of the disease in a controlled setting. Although some human challenge studies had previously been conducted for diseases like cholera, dengue, influenza and malaria, they were generally limited to well-understood infectious strains known to cause mild disease (Weijer 2024). This was not the case for COVID-19.

Debates also arose about balancing opportunities to conduct COVID-19 clinical research with the urgent need to prioritise clinical care for patients (Hashem et al. 2020). What is more, few research ethics committees already had internal policies to guide activities during public health emergencies, so most had to modify existing procedures or develop new ones and had no time to evaluate those changes (Salamanca-Buentello et al. 2024). The PREPARED consortium identified 236 new sets of ethics guidelines for the COVID-19 pandemic alone (see Chap. 4). Amid this abundance of existing codes and guidelines for research ethics and integrity, it is reasonable to ask if we need another – and, if so, why.

4 What is Different About the PREPARED Code?

This book is about yet more ethics guidance, the PREPARED Code: A Global Code of Conduct for Research During Pandemics. But it is ethics guidance that stands out in five ways. The PREPARED Code:

1. focuses on one very specific area, *research during pandemics*, which makes it easy for researchers from any discipline to find guidance quickly and easily should the need arise
2. is short and jargon-free (unlike most ethics guidance), thereby enhancing accessibility
3. is based on significant, global research undertaken in nine languages to identify real-world challenges during pandemics
4. is values-driven to motivate users to understand why they should comply with the guidance articles, for instance the need to take care that additional responsibilities during pandemics are distributed fairly and in a way that does not exacerbate existing inequities
5. combines research ethics and research integrity advice to stop the silo-building that divides these two sister disciplines and results in their generally being addressed as if they were separate entities rather than two sides of the same coin, both concerned with doing the right thing in research.

The approach used for the development of the PREPARED Code was previously applied in the development of The TRUST Code: A Global Code of Conduct for Equitable Research Partnerships (TRUST 2018). One could venture to call the TRUST Code, launched in 2018, the most successful ethics code of the past ten years, given its rapid endorsement by high-profile adopters from around the world, including research funders, the European Commission, the European & Developing Countries Clinical Trials Partnership, the Dutch and Polish governments, and the publishers *Nature* and Sage across their entire portfolios (European Commission 2018; Nature Medicine 2023).

This book introduces the PREPARED Code as follows:

Chapter 2 presents the PREPARED Code in its final form, the end product of a long and in-depth development process.

Chapter 3 focuses on explaining and justifying the guiding rationale of the development process. It explores the conceptual foundation of the PREPARED Code, including why a risk-based and values-driven approach was taken, and why research ethics and research integrity are combined in one code.

Chapter 4 describes the broad research foundation on which the PREPARED Code was built, including literature reviews of the challenges related to COVID-19 in nine languages, scoping reviews on other epidemics and pandemics, special group reports for vulnerable persons, a human rights report, and the identification and analysis of related ethics guidance documents.

Chapter 5 outlines the methods that were employed at each stage of development to elucidate how the code was built, from the identification of the risks to research ethics and integrity during pandemics to the iterative and broad consultation applied in refining the resulting code.

Chapter 6 describes the implementation support developed to help people understand the pandemic-related ethics and integrity challenges and how to apply the PREPARED Code, including tools like a specially designed app that guides the learner through a wide range of relevant case studies.

Chapter 7 synthesizes the learning from a broad range of activities to develop a code of conduct to guide research during pandemics. The chapter includes our recommendations for future developers of ethics codes to help ensure effectiveness and credibility.

References

Barroga, E., Matanguihan, G.J.: Fundamental shifts in research, ethics and peer review in the era of the COVID-19 pandemic. J. Korean Med. Sci. **35**(45), e395 (2020). https://doi.org/10.3346/jkms.2020.35.e395

Bauchner, H., Fontanarosa, P.B., Golub, R.M.: Editorial evaluation and peer review during a pandemic: how journals maintain standards. JAMA **324**(5), 453–454 (2020). https://doi.org/10.1001/jama.2020.11764

Bratan, T., Aichinger, H., Brkic, N., et al.: Impact of the COVID-19 pandemic on ongoing health research: an ad hoc survey among investigators in Germany. BMJ Open **11**(12), e049086 (2021). https://doi.org/10.1136/bmjopen-2021-049086

Brierley, L.: Lessons from the influx of preprints during the early COVID-19 pandemic. Lancet Planet. Health **5**(3), e115-117 (2021). https://doi.org/10.1016/S2542-5196(21)00011-5

Dinis-Oliveira, R.J.: COVID-19 research: pandemic versus "paperdemic", integrity, values and risks of the "speed science." Forensic Sci. Res. **5**(2), 174–187 (2020). https://doi.org/10.1080/20961790.2020.1767754

Doyle, J.M., Morone, N.E., Proulx, C.N., et al.: The impact of the COVID-19 pandemic on underrepresented early-career PhD and physician scientists. J. Clin. Transl. Sci. **5**(1), e174 (2021). https://doi.org/10.1017/cts.2021.851

Economist: Speeding up science during the pandemic, 9 May 2020. https://www.economist.com/leaders/2020/05/09/speeding-up-science-during-the-pandemic. Accessed 9 Feb 2025

European Commission: Success stories: a global code of conduct to counter ethics dumping, 27 June 2018. https://projects.research-and-innovation.ec.europa.eu/en/projects/success-stories/all/global-code-conduct-counter-ethics-dumping. Accessed 9 Feb 2025

Fraser, N., Brierley, L., Dey, G., et al.: The evolving role of preprints in the dissemination of COVID-19 research and their impact on the science communication landscape. PLoS Biol. **19**(4), e3000959 (2021). https://doi.org/10.1371/journal.pbio.3000959

Gaudino, M., Arvind, V., Hameed, I., et al.: Effects of the COVID-19 pandemic on active non-COVID clinical trials. J. Am. Coll. Cardiol. **76**(13), 1605–1606 (2020). https://doi.org/10.1016/j.jacc.2020.07.051

Harper, L., Kalfa, N., Beckers, G.M., et al.: The impact of COVID-19 on research. J. Pediatr. Urol. **16**(5), 715–716 (2020). https://doi.org/10.1016/j.jpurol.2020.07.002

Hashem, H., Abufaraj, M., Tbakhi, A., Sultan, I.: Obstacles and considerations related to clinical trial research during the COVID-19 pandemic. Front. Med. **7**, 598038 (2020). https://doi.org/10.3389/fmed.2020.598038

HHS: International compilation of human research standards. Office for Human Research Protections, US Department of Health and Human Services (2024). https://www.hhs.gov/ohrp/international/compilation-human-research-standards/index.html. Accessed 9 Feb 2025

Heo, S., Chan, A.Y., Diaz Peralta, P., et al.: Impacts of the COVID-19 pandemic on scientists' productivity in science, technology, engineering, mathematics (STEM), and medicine fields. Humanit. Soc. Sci. Commun. **9**(1), 1–11 (2022). https://doi.org/10.1057/s41599-022-01466-0

Independent Panel: COVID-19: make it the last pandemic. Independent Panel for Pandemic Preparedness & Response (2021). https://theindependentpanel.org/wp-content/uploads/2021/05/COVID-19-Make-it-the-Last-Pandemic_final.pdf. Accessed 9 Feb 2025

Kebenei, E.K., Cheruiyot, D., Msee, G.C., et al.: The impact of Covid-19 pandemic on the research portfolio and approval turnaround time at the Kenya medical research institute. Indian J. Med. Ethics **9**(2), 109–114 (2024). https://doi.org/10.20529/IJME.2024.013

Law, E., Smith, I.: Ethical and informative trials: how the COVID-19 experience can help to improve clinical trial design. Res. Ethics **20**(4), 764–779 (2024). https://doi.org/10.1177/17470161241261768

Maison, D., Jaworska, D., Adamczyk, D., Affeltowicz, D.: The challenges arising from the COVID-19 pandemic and the way people deal with them: a qualitative longitudinal study. PLoS ONE **16**(10), e0258133 (2021). https://doi.org/10.1371/journal.pone.0258133

Medicine, N.: Editorial: striving for equitable partnerships in health research. Nat. Med. **29**, 2667–2668 (2023). https://doi.org/10.1038/s41591-023-02680-2

Omary, M.B., Eswaraka, J., Kimball, S.D., et al.: The COVID-19 pandemic and research shutdown: staying safe and productive. J. Clin. Investig. **130**, 2745–2748 (2020). https://doi.org/10.1172/JCI138646

Radecki, J., Schonfeld, R.: The impacts of COVID-19 on the research enterprise: a landscape review. Ithaka S+R (2020). https://sr.ithaka.org/wp-content/uploads/2020/10/SR-Report-Impacts-of-COVID-19-on-the-Research-Enterprise102620-1.pdf. Accessed 9 Feb 2025

Resnik, D.B.: Research ethics timeline (1932–present) (2018). https://www.aum.edu/wp-content/uploads/2022/09/Research-Ethics-Timeline.pdf. Accessed 9 Feb 2025

Salamanca-Buentello, F., Katz, R., Silva, D.S., et al.: Research ethics review during the COVID-19 pandemic: an international study. PLoS ONE **19**(4), e0292512 (2024). https://doi.org/10.1371/journal.pone.0292512

Saxena, A., Bouvier, P.A., Shamsi-Gooshki, E., et al.: WHO guidance on ethics in outbreaks and the COVID-19 pandemic: a critical appraisal. J. Med. Ethics **47**(6), 367–373 (2021). https://doi.org/10.1136/medethics-2020-106959

Sharma, R.K., Teng, A., Asirot, M.G., et al.: Challenges and opportunities in conducting research with older adults with dementia during COVID-19 and beyond. J. Am. Geriatr. Soc. **70**(5), 1306–1313 (2022). https://doi.org/10.1111/jgs.17750

Shawrav, M.M.: How the pandemic shaped the research environment. European Research Council Magazine, 28 March (2022). https://erc.europa.eu/news-events/magazine/how-pandemic-shaped-research-environment? Accessed 9 Feb 2025

Solbakk, J.H., Bentzen, H.B., Holm, S., et al.: Back to WHAT? The role of research ethics in pandemic times. Med. Health Care Philos. **24**, 3–20 (2021). https://doi.org/10.1007/s11019-020-09984-x

Tegli, J.K.: Seeking retrospective approval for a study in resource-constrained Liberia. In: Schroeder, D., Cook, J., Hirsch, F., Fenet, S., Muthuswamy, V. (eds.) Ethics Dumping. SpringerBriefs in Research and Innovation Governance, pp. 115–119. Springer, Cham (2018). https://doi.org/10.1007/978-3-319-64731-9_14

TRUST: The TRUST code: a global code of conduct for equitable research partnerships (2018). https://doi.org/10.48508/GCC/2018.05. Accessed 5 Feb 2025

Weijer, C.: COVID-19 human challenge trials and randomized controlled trials: lessons for the next pandemic. Res. Ethics **20**(4), 636–649 (2024). https://doi.org/10.1177/17470161231223594

WHO: A healthy return: Investment case for a sustainably financed WHO. World Health Organization, Geneva (2022). https://www.who.int/about/funding/invest-in-who/investment-case-2.0. Accessed 9 Feb 2025

Xue, J.Z., Smietana, K., Poda, P., et al.: Clinical trial recovery from COVID-19 disruption. Nat. Rev. Drug Discov. **19**(10), 662–663 (2020). https://doi.org/10.1038/d41573-020-00150-9

Zhaksylyk, A., Zimba, O., Yessirkepov, M., Kocyigit, B.F.: Research integrity: where we are and where we are heading. J. Korean Med. Sci. **38**(47), e405 (2023). https://doi.org/10.3346/jkms.2023.38.e405

Open Access This chapter is licensed under the terms of the Creative Commons Attribution 4.0 International License (http://creativecommons.org/licenses/by/4.0/), which permits use, sharing, adaptation, distribution and reproduction in any medium or format, as long as you give appropriate credit to the original author(s) and the source, provide a link to the Creative Commons license and indicate if changes were made.

The images or other third party material in this chapter are included in the chapter's Creative Commons license, unless indicated otherwise in a credit line to the material. If material is not included in the chapter's Creative Commons license and your intended use is not permitted by statutory regulation or exceeds the permitted use, you will need to obtain permission directly from the copyright holder.

The PREPARED Code: A Global Code of Conduct for Research During Pandemics

Doris Schroeder[1,2](✉) and David Robinson[3]

[1] School of Law, UCLan Cyprus, Larnaca, Cyprus
DSchroeder@uclan.ac.uk
[2] Centre for Professional Ethics, University of Central Lancashire, Preston, UK
[3] DaveRob Design and Illustration, Preston, UK

Abstract. This chapter presents the 27 articles of the PREPARED Code: A Global Code of Conduct for Research during Pandemics. Illustrative diagrams capture essential features of its development (e.g. the geographical distribution of the authors) and structure (e.g. that it is built around fairness, respect, care and honesty).

Keywords: Research ethics · research integrity · pandemics

1 Introduction

Research ethics and integrity challenges during pandemics are not unique, but they are vastly magnified during crises.

The PREPARED Code for researchers, research ethics committees and research integrity offices applies throughout a pandemic. The code was developed by an international consortium and is based on research undertaken in English, Chinese, French, German, Hindi, Japanese, Korean, Russian and Spanish. It was refined through a human rights analysis and extensive consultation with stakeholders. Input from marginalized populations was obtained at every stage.

The PREPARED Code:

- Respects the Declaration of Helsinki as the primary source of research ethics guidance during pandemics.
- Provides support across all research disciplines.
- Presents concise statements in clear language to encourage access.
- Combines guidance on research ethics and integrity.
- Complements the TRUST Code and the European Code of Conduct for Research Integrity, because the risks of inequitable research and breaches of research integrity can increase during a crisis.
- Links each guidance article to the values of fairness, respect, care and honesty.

VISION: Pandemic research should be trustworthy and the results accessible to all.

Citing suggestion for the PREPARED Code: PREPARED (2025) The PREPARED Code – A Global Code of Conduct for Research during Pandemics, https://preparedcode.uclancyprus.ac.cy/

Fairness

Article 1

Data and scientific insights about new infectious agents should be quality controlled and **shared** as swiftly as possible with the scientific community and other stakeholders, without prejudice to the sharer.

Article 2

Research coordination and cooperation are essential to avoid the unnecessary duplication of studies, which could place unfair burdens on participants and waste time and resources.

Article 3

A fair plan for **access to the benefits** of pandemic research should be agreed early on in any project, in collaboration with stakeholders.

Article 4

Where possible, **community engagement** should be continued or even increased during a pandemic, to address the most pressing needs of communities and to help maintain trust in science.

Article 5

Vulnerabilities increase during pandemics. Where possible, research approaches should be adapted to ensure the **ethical inclusion of persons in vulnerable situations** – with adequate protections – rather than adopting patronizing or convenience exclusions.

Article 6

Research teams should share the **additional responsibilities** associated with a pandemic fairly among their members to avoid exacerbating existing inequalities.

Respect

Article 7

Research ethics committee (REC) guidance and approval should be sought and respected at all times, including during pandemics. RECs should **expedite the evaluation** of research proposals that address urgent societal needs without compromising rigorous ethical standards.

Article 8

Community researchers are part of the research team and should be treated and respected as researchers, including during pandemics.

Article 9

The urgent need to conduct research can never be an excuse for putting pressure on potential research participants or their proxies to make hasty decisions about their involvement in a study. **Genuine informed consent needs time.**

Article 10

Changes to the process of seeking **informed consent** must not be allowed to compromise potential participants' understanding of a research project. This includes ensuring that research participants do not mistake research for treatment ('therapeutic misconception'), especially when healthcare staff rather than researchers seek consent.

Article 11

The informed consent process should explain the study **risks** and benefits fully and clearly in terms of what is known, what is **uncertain** and what is unknown.

Article 12

During pandemics, all those involved in the research cycle should strive for **respectful engagement** with each other in the spirit of equitable and collaborative problem-solving.

Article 13

Researchers must always use **respectful language** when communicating through the press or the media, even when under pressure.

Care

Article 14

Research must not compromise **public health responses**. In particular, the involvement of clinical staff in research should not affect patient care negatively.

Article 15

Especially during pandemics, researchers who handle potentially infectious **biological materials** should be adequately **trained** and equipped to safeguard public health.

Article 16

Researchers should keep in mind how pandemic conditions may affect all stakeholders in a study (participants, healthcare staff, support staff etc.) and take appropriate measures to **ease any additional burdens**.

Article 17

When **research is prioritized during a pandemic**, research participants in ongoing studies must not be left worse off than before they joined their original study.

Article 18

Where research participants depend on research studies for access to medication and services, **study modifications** during pandemics need to be managed responsibly to ensure that their lives and health are not endangered.

Article 19

During pandemics, studies involving **healthy volunteers** in which novel compounds are administered to humans or no rescue therapy is available should only be started if space in intensive care units is assured for the needs of healthy volunteers, as well as for all patients in routine care.

Article 20

In the context of uncertainty, researchers should **review their study protocols regularly** to ensure that new findings are taken into account as they emerge.

Article 21

During pandemics, researchers may experience a **heightened risk of hostility** and related safety and security concerns. Research ethics committees should check that risk management plans are in place.

Honesty

Article 22

It is vital that researchers uphold the **highest standards of research integrity**, even when under significant pressure, to ensure the reliability of pandemic research results and to maintain public trust in science.

Article 23

Participants and research ethics committees should be **promptly** and fully informed about changes in the risks or burdens of participation in clinical research if **new, relevant information** becomes available during a trial.

Article 24

Existing regulatory requirements for the **secondary use** of personal data and biological materials must prevail during pandemics, unless an explicit exception has been enacted.

Article 25

Researchers should actively **support** rigorous, **fast-track scientific review** to help combat the erosion of good science during pandemics. They should also support quality control mechanisms for open communication channels such as pre-print servers or social media.

Article 26

Researchers should answer **publishers' research ethics questions** in full, even in rapid review submissions.

Article 27

In **public communications**, researchers should ensure that the scientific information presented is reliable. They should be clear about study limitations and avoid exaggeration, sensationalism and deception.

The code was drafted as part of the PREPARED project under the lead author Prof. Doris Schroeder. The code was developed for pandemics, but may also be useful for epidemics and public health emergencies of international concern.

2 The PREPARED Code in Diagrams

The following diagrams present information on the PREPARED Code visually.

The co-authors of the PREPARED Code are based in 22 countries, on five continents (Fig. 1). Most of the partner teams are led by women.

The PREPARED Code is one of six major outputs from the PREPARED team, all relevant to maintaining research ethics and integrity in a global crisis (Fig. 2). How other outputs facilitate the use of the PREPARED Code is explained in Chap. 6.

The PREPARED Code is currently available in 13 languages, covering the most common official languages in the world (except Russian), namely: English, Spanish, Chinese, French, Arabic and Portuguese (Fig. 3).

The PREPARED Code is the sister code of the TRUST Code (TRUST 2018) and the San Code of Research Ethics (SASI 2017). The moral framework of all three aligns around fairness, respect, care and honesty. A fourth sister code will be released later in 2025, covering research ethics and research integrity in fragile settings, such as conflict zones (Fig. 4).

Fig. 1. PREPARED Code authors: the geography

Fig. 2. One of six major outputs

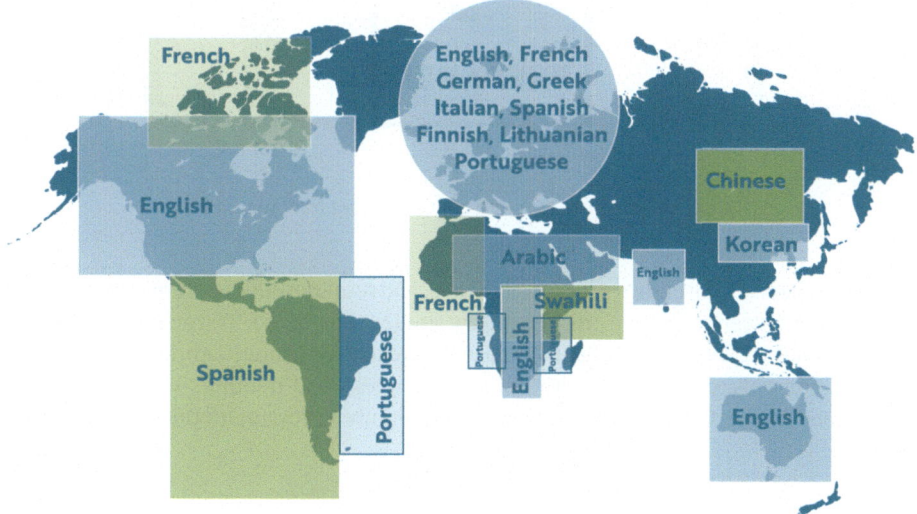

Fig. 3. Translations

Fig. 4. Sister codes

Fig. 5. Marrying research ethics and research integrity

Researchers are the code's primary audience, and research ethics committee members and research integrity officers its secondary audience (Fig. 5). As noted in Chap. 3, research ethics and research integrity are usually treated as discrete entities with separate guidance, journals, conferences, professional networks and training. The PREPARED Code is one of the few guidance documents to combine research ethics and integrity articles.

References

SASI: San Code of Research Ethics. South African San Institute, Kimberley (2017). https://www.globalcodeofconduct.org/wp-content/uploads/2018/04/San-Code-of-RESEARCH-Ethics-Booklet_English.pdf. Accessed 19 Feb 2025

TRUST: The TRUST code: a global code of conduct for equitable research partnerships (2018). https://doi.org/10.48508/GCC/2018.05

Open Access This chapter is licensed under the terms of the Creative Commons Attribution 4.0 International License (http://creativecommons.org/licenses/by/4.0/), which permits use, sharing, adaptation, distribution and reproduction in any medium or format, as long as you give appropriate credit to the original author(s) and the source, provide a link to the Creative Commons license and indicate if changes were made.

The images or other third party material in this chapter are included in the chapter's Creative Commons license, unless indicated otherwise in a credit line to the material. If material is not included in the chapter's Creative Commons license and your intended use is not permitted by statutory regulation or exceeds the permitted use, you will need to obtain permission directly from the copyright holder.

Ensuring Effectiveness and Credibility: The Conceptual Foundation of the PREPARED Code

Kate Chatfield[1](✉), Doris Schroeder[2], Eugenijus Gefenas[3], Vilma Lukaševičienė[3], Kalle Videnoja[4], Emma Law[5], Joyce Adhiambo Odhiambo[6], and Joshua Kimani[6]

[1] Centre for Professional Ethics, University of Central Lancashire, Preston, UK
kchatfield@uclan.ac.uk
[2] School of Law, UCLan Cyprus; Centre for Professional Ethics, University of Central Lancashire, Preston, UK
[3] Faculty of Medicine, Vilnius University, Vilnius, Lithuania
[4] Finnish National Board On Research Integrity (TENK), Helsinki, Finland
[5] PROTAS, London, UK
[6] Partners for Health and Development in Africa, Nairobi, Kenya

Abstract. This chapter explains the conceptual foundations of the PREPARED Code, which together provide the credibility required to justify adding yet another ethics code to the thousands that already exist. The code is built on real-world risks identified in nine languages rather than, for instance, on drafters' expertise, thereby making it as precisely honed an instrument as possible to cope with the real-world ethics and integrity challenges experienced during a pandemic. The code is values-driven, focused on the values of fairness, respect, care and honesty, to harness the motivational power of moral values and to provide an easily understandable, globally applicable moral framework. Unlike most other ethics codes, the PREPARED Code unites research ethics and research integrity guidance into one, to ensure that a culture of integrity rather than a box-ticking mentality is fostered. The short, jargon-free code text addresses all research disciplines and, most importantly, it is based on extensive input from a wide range of stakeholders, including highly marginalised populations, to ensure that it is fit for purpose.

Keywords: Research ethics · research integrity · pandemic ethics · ethics codes

1 Introduction

Drafting an ethics code is something anyone can attempt, but crafting a code that is effective and credible demands thoughtful attention to a variety of factors. An effective ethics code serves not just as a set of rules for guiding conduct, but also as a practical and inspirational tool for fostering a culture of integrity. Additionally, those who use an ethics code need to trust that it is fit for purpose and can serve the interests of all stakeholders. How can this be ensured?

While the visible *product* of development, the resultant ethics code, becomes the familiar tool, it is the behind-the-scenes *process* of code development that confers credibility (Messikomer and Cirka 2010). Thus, the short and concise PREPARED Code presented in Chap. 2 may look as though it was quick and easy to develop, but it is underpinned by a significant body of work that was guided by a clear and coherent rationale.

This chapter focuses on explaining and justifying the guiding rationale for the development process. In other words, it explores the conceptual foundations of the PREPARED Code, addressing "why" rather than "how" questions, including:

- *Why is the PREPARED Code built on real-world risks?* Section 2 examines different approaches to developing ethics codes, including top-down and bottom-up methods, and explains why a bottom-up, risk-based approach was selected.
- *Why is the PREPARED Code values-driven?* Section 3 explains the advantages of values-driven moral frameworks over mere rules-based systems. It also addresses why the moral values of fairness, respect, care and honesty serve as the pillars of the code.
- *Why are research ethics and research integrity integrated in a unified code?* Section 4 outlines the differences, commonalities and complex interplay between research integrity and research ethics. It also clarifies the advantages of combining research ethics and research integrity in one code, especially during a pandemic, rather than regarding them as separate silos.
- *Why does the code take a broad and inclusive approach to development?* Section 5 examines why the wider social, economic and cultural contexts in which pandemics occur were considered, stressing the importance of addressing intersecting vulnerabilities. Additionally, the imperative of inclusivity in the code's development was emphasised, ensuring that the voices of *all* stakeholders, especially marginalised populations, were heard and their views reflected in the code.
- *Why is the PREPARED Code short, jargon-free, multidisciplinary and focused?* The chapter ends with an explanation of the relationship of the PREPARED Code to its sister code, the TRUST Code (TRUST 2018), to show why a short, jargon-free and focused code that addresses all academic disciplines can succeed among a proliferation of ethics guidance.

2 Approaches to the Development of Ethics Codes

Various approaches have been utilised to create ethics codes, each employing distinct methodologies that come with their own strengths and limitations. In this section we consider the main methods that can be used and explain the decision to use a risk-based approach plus extensive consultations for the development of the PREPARED Code. The four approaches are summarised in Fig. 1.

Approaches to drafting ethics codes can be broadly distinguished as "top-down" or "bottom-up". Of the top-down approaches, the most common and well-known method is the *drafter-based approach*, which relies upon the drafters of an ethics code or legal document to identify the main content (for example, the challenges to be addressed) as well as the underlying principles or values.

This approach is likely to have been used for the UN (1948) Universal Declaration of Human Rights. For instance, when South African Prime Minister Jan Christiaan Smuts suggested lines for the preamble in 1945, he referred to "fundamental human rights"

Fig. 1. Possible approaches to ethics code development

and "the sanctity and ultimate value of human personality". Yet the drafters changed "sanctity" to "dignity" to achieve the broad consensus required (Tiedemann 2006).

Another top-down approach to developing an ethics code can be termed the *code-based approach*, as it involves analysis of existing ethics codes and documents to decide what is relevant to the development of new guidance. This approach is often used alongside a drafter-based method, with a code-based version being produced as a starting point that the drafters subsequently revise.

For both drafter-based and code-based methods, the development of ethics guidance documents is centralised and reliant upon representatives of international bodies, professional organisations and experts having the relevant knowledge and experience. Consequently, while top-down ethics codes are representative of the opinions of experts, especially those from the research governance community, they might not incorporate the opinions and experiences of other important stakeholders. Hence, the major shortcoming of these two top-down approaches is that they rely upon the viewpoints of a group of drafters. In the case of code-based approaches, this is because (at least some of) the codes and guidance used to inform the new code will themselves have been developed using the drafter-based approach.

By contrast, ethics codes can also be developed via bottom-up approaches that proactively seek to represent the experiences and opinions of a wide range of stakeholders. For instance, citizen-driven approaches are increasingly used across a variety of domains, such as innovation and policymaking (Paleco et al. 2021; Huttunen et al. 2022). However, a purely citizen-driven approach to development was not considered suitable for the PREPARED Code because very few citizens will have experience of research ethics and research integrity challenges during pandemics. To be effective, ethics codes must address real-world challenges rather than those of a hypothetical or speculative nature.

Thus, an alternative bottom-up approach was taken, involving considerable effort to reveal real-world research ethics and research integrity challenges and corresponding risks arising in the context of pandemic research. Rather than consulting experts, existing codes or citizens to identify the matters to be included in the code, the PREPARED team identified real-world research ethics and research integrity issues that had actually occurred from published materials in nine languages (English, Chinese, French, German, Hindi, Japanese, Korean, Russian and Spanish), as well as from studies with certain groups who were likely to have been disadvantaged during the COVID-19 pandemic (in particular disabled people, health and social care workers, people on the poverty line and women). (For details, see Chap. 4.) This risk-based approach to the identification of matters that raise ethical issues for research during pandemics meant that the resultant code was evidence-informed, and this helped to ensure its pragmatic value because it was grounded in real-world experience.

The risk-based approach has become prominent across a number of policy domains, including public health, finance and disaster management, as a method to allocate resources efficiently and to address high-priority risks (Rothstein et al. 2013). This is because risk-based approaches can help to align governance goals with levels of risk, supporting rational decision-making and institutional accountability (Graham 2010). Unlike drafter-based approaches, which often rely on subjective expertise, the risk-based method employs systematic evaluations, enabling more objective, data-informed decisions (Black and Baldwin 2010). The approach also integrates the precautionary principle, which advocates preventive action in the face of scientific uncertainty (Khodadadyan et al. 2021).

It is important to note, however, that the four approaches are not mutually exclusive. For instance, the recent revision of the World Medical Association (WMA) Declaration of Helsinki (WMA 2024b), which outlines the ethical principles for medical research involving human participants, combined a drafter-based approach with very significant external consultations, thereby using elements of the citizen-driven approach. The workgroup entrusted with the ninth revision of the declaration included representatives from 19 countries and invited advisers with expertise in bioethics. Over the course of 30 months, the workgroup received feedback from partners, including WMA members, who provided comments on the draft text during two public comment periods. Additionally, eight regional meetings were held across all WMA regions. Ultimately, the refined draft of the declaration was unanimously adopted by the 50 delegates at the General Assembly of the WMA in October 2024, marking 60 years since the original declaration was adopted (Resneck 2024; WMA 2024a).

For the PREPARED Code, risk-based analysis *drove* the development. However, the code authors also included elements from the code-based and citizen-based approaches. (See Chap. 5 for the extensive, including public, consultations and Chap. 4 for the gap analysis using existing ethics codes relevant to pandemics.) Nevertheless, we believe that the risk-based approach should precede all other approaches to produce an effective, evidence-informed code. That is because risk-based approaches are closest to real-world challenges.

In the next section we explain why the PREPARED Code is values-driven, and why the moral values of fairness, respect, care and honesty serve as the pillars of this code.

3 A Values-Driven Code

Just as there are various approaches that could have been taken to code development, there are also various ethics frameworks that might have been adopted. For instance, ethics codes and guidance documents can be rules-based, principles-based or human-rights-based, with guidance statements designed in alignment with certain moral rules, moral principles or human rights. Box 3.1 indicates the distinctions that were drawn between rules, principles, values and virtues for the development of the TRUST Code (TRUST 2018) and the PREPARED Code. Other interpretations are possible.

> **Box 3.1 Rules, Principles, Values and Virtues**
> - *Rules* are universal directives that specify what is permissible and what is forbidden. Actions are deemed "morally correct" if they align with specific moral rules, such as "do not kill", regardless of the outcome.
> - *Principles* are behavioural rules for action in specific situations. When the principle is known, it is clear what to do. For instance, *in dubio pro reo* is a principle which requires that people be treated as innocent until proven guilty.
> - *Values* can be mathematical; they can refer to things people desire (i.e. value), e.g. money or glory; or they can be moral values, i.e. guidance for doing the right thing, such as trying to be fair.
> - *Virtues* and values often refer to the same moral entity, e.g. fairness, but in the case of virtues, they are ingrained solidly in a person's character, which does not have to be the case for values.

The decision to adopt a values-based framework for the PREPARED Code was taken for the reasons outlined below.

We believe that there is a limit to the usefulness of principles- or rules-based research ethics and research integrity approaches and frameworks. Ethicists have long contended that systems based solely on rules or principles, without incorporating agent-centred virtues or values, can be deeply problematic because they disconnect moral behaviour from the individual's character and intrinsic motivations (Johnsson et al. 2014). Rules alone do not motivate to action (Dawson 1994), especially if the potential outcomes do not appear to be right or fair.

When set solely within a compliance-based framework, research ethics and research integrity can seem like a box-ticking exercise, reducing ethical responsibility to a mere checklist (Pennock and O'Rourke 2017). Thus, of recent years, particularly in the domain of research integrity, there has been a notable shift towards the promotion of virtue ethics (Banks 2018), with the assumption that this agent-centred approach will serve to engender moral character and a greater sense of personal responsibility in researchers (Mitchell 2015). An agent-based ethics framework offers an alternative to exclusively rules- or principles-based methods. In line with this shift, the underpinning ethics framework for the PREPARED Code promotes an agent-centred approach, albeit via moral values rather than virtues.

We accept that the nature of the relationship between moral values and virtues is a matter of debate, but our stance assumes that a values approach and a virtues one might be regarded as different points along the same trajectory (Chatfield and Law 2024). For instance, if a person cultivates the value of honesty, it might eventually become a virtue of that person, virtues being embedded moral values. In the main, the development of moral virtues relies upon habit. Eventually, through much practice, habits become character traits of the virtuous individual (Aristotle 2009). While the development of the virtuous researcher is an admirable goal, it might – given that it takes years of dedication and practice – feel exclusionist to a young researcher. Alternatively, a values approach can tap into (hopefully) existing moral values that resonate with even a novice researcher.

Significantly for research ethics and research integrity, a defining characteristic of personal values is their motivational power. This applies especially to values with explicit moral significance, which are often regarded as the most important in moral motivation (Schwartz 2012). People hold their moral values in high esteem, allowing them to shape behaviour and decision-making profoundly (Schroeder et al. 2018). They guide decision-making, inclining us towards one course of action over another (Ogletree 2004). Moral values are especially important for directing ethical choices. For example, holding fairness as a core value *motivates* us to treat people fairly and incorporate fairness into our decision-making.

The most prominent example of a values-based approach in research ethics is found in the TRUST Code (TRUST 2018), which is grounded in the values of fairness, respect, care and honesty. While other moral frameworks are regularly criticised as overly Western or Anglo-Saxon, especially systems that accord "autonomy" intrinsic value (Varelius 2006), the TRUST Code has been adopted around the world. It is used globally to guide equitable research partnerships (Chatfield and Law 2024) and its widespread adoption may be attributed to three main factors:

1. The four values of fairness, respect, care and honesty are straightforward and accessible, requiring no technical expertise to be understood.
2. They were identified as components of a moral framework through a bottom-up process, avoiding bias toward high-income country perspectives (Schroeder et al. 2018).
3. The framework was developed by a diverse global team, including representatives from vulnerable populations (ibid).

Nevertheless, while the TRUST values had resonated globally, their applicability to the PREPARED Code could not be assumed. Prior to the identification of the wide-scale research ethics and research integrity risks encountered during pandemics, alignment of the risks with the TRUST values was purely a matter of speculation. Nevertheless, as described in Chap. 5, it soon became clear, during the process of risk analysis for the PREPARED Code, that *all* of the identified pandemic-related risks for research ethics and research integrity could be aligned with at least one of the four TRUST values. In other words, the identified breaches of research ethics and research integrity that emerged or were exacerbated during pandemics could all be associated with lapses or failures in fairness, respect, care and/or honesty.

Explanation of the four values within the context of research is articulated elsewhere (Schroeder and Chatfield 2018; Schroeder et al. 2018; Chatfield and Law 2024), but it is easy to imagine why fairness, respect, care and honesty are also important in the context of a pandemic as illustrated in Fig. 2.

While the values of fairness, respect, care and honesty can provide a moral framework to guide ethical research during a pandemic, it is not always easy to understand how values should be applied in specific situations. Thus, both the TRUST Code (TRUST 2018) and the PREPARED Code include guidance articles that help to operationalise the values. The PREPARED Code lists 27 articles: six for fairness, seven for respect, eight for care and six for honesty. (The process of alignment of articles and values is described in Chap. 5).

This linkage makes it easier for PREPARED Code users to understand the relationship between their values and the action-guiding articles. For instance, by attempting to continue community engagement during a pandemic, researchers can be sure that they are enacting fairness.

Fig. 2. A world of fairness, respect, care and honesty during pandemics

Now that the reasons for adopting a risk-based and values-driven approach to code development have been explained, the next section explains why the PREPARED team integrated research ethics and research integrity in a unified code.

4 Research Ethics and Research Integrity in the Same Code?

The PREPARED Code is different from most other codes or guidelines for responsible research in that it addresses matters of both research ethics and research integrity in one code. This is highly unusual, as they are more ordinarily viewed as distinct concepts that require discrete regulation (Kolstoe and Pugh 2023). To understand why this is so, it is helpful to consider the provenance of research ethics and research integrity.

It is often repeated that research ethics was "born in scandal and reared in protectionism" (Levine 1988; see also Reverby 2012; Dhai 2014), thanks to shocking revelations about unethical research that shaped attitudes, guidance and legislation around research ethics significantly. For instance, the Nuremberg Code was developed in response to the horrific medical experiments conducted by Nazi doctors during World War II (Annas and Grodin 2008). It became the first internationally recognised guidance document in research ethics, emphasising the requirement for informed consent and the duty to protect participants from harm.

Nowadays the term "research ethics", in the broadest sense, is applied to all issues of a moral nature that are associated with the planning, conduct, dissemination and impacts of research. Additionally, beyond the participation of humans, research ethics also governs aspects such as harm to animals or the environment, dual use (the use of research results by both civilians and the military), misuse, impacts upon societies (e.g. from AI technologies) and impacts upon communities (e.g. ethics dumping, which is the offshoring of unethical research from higher-income to lower-income regions (Schroeder et al. 2018)).

Similarly, over the past thirty years a growing level of concern around the integrity of research has been driven by scandals including unsafe practices, the falsification of data and concerns about the reliability of research outputs. For instance, in the late 2000s, news emerged about scientific misconduct involving researcher Anil Potti, who fabricated and falsified data in cancer studies in the United States. Potti's false claim to have developed a method for personalising cancer treatments (Ince 2011) revealed a lack of independent validation of data and insufficient oversight of clinical trials (Kurzrock et al. 2014).

Around the same time, revelations about Hwang Woo Suk's fabrication of results in cloning and stem cell research prompted international guidelines emphasising transparency, reproducibility and peer review (Franzen 2016; Wilson 2020).

Thus, instances of research misconduct like these research ethics scandals led to the development of governance procedures.

Research misconduct does not just lead to lost time, lost resources and damaged reputations (Committee on Science, Engineering, and Public Policy 2009:18).

> The consequences of research misconduct can be severe including preventable illness or the loss of human life due to misinformation in the literature or continued citing of retracted work. (Imperial College London n.d.)

Today, the term "research integrity" is associated with "the conduct of research" in ways that "promote trust and confidence" in "all aspects of research" (UKRIO 2023). When researchers conduct their research with integrity, this enables the global research community and society to have confidence and trust in the methods *and* the findings of the research (Dove 2024).

While being "born in scandal" (Levine 1988) applies to both fields, research ethics and research integrity have become generally regarded as discrete entities with separate journals, separate conferences, separate professional networks and often separate training units (Chatfield and Law 2024). Table 1 lists some examples of commonly perceived differences between research ethics and research integrity.

Table 1. Summary overview of perceived differences between research ethics and research integrity

	Research Ethics	Research Integrity
Main purpose	Protection of rights and welfare of participants	Trustworthiness of science
How governed?	Research ethics committee approvals and monitoring	Research integrity institutional frameworks, boards and officers
Training	Process focus, with rules and principles	Agent focus, with virtues and values
Timing	Prospective assessment (before a study)	Retrospective assessment (after a study)
Research quality	The focus not on scientific quality	The primary focus on scientific quality

The examples above serve to illustrate why the two fields are often governed separately, but the differences listed in Table 3.1 are oversimplified. In practice, the lines of division between research ethics and research integrity are often blurred. Taking research quality as an example, an ongoing question of research ethics committees has been whether bad science equals bad ethics (Dawson and Yentis 2007). As became starkly apparent during the COVID-19 pandemic, conducting trials whose ultimate failure to answer the research question is foreseeable (e.g. because the trials are not large enough to provide a definitive result) is not only a waste of resources but also a breach of research participants' trust and a violation of research ethics (Law and Smith 2024).

Even though there may appear to be clear differences between research ethics and research integrity, there is also a significant overlap. Overlapping issues include conflicts of interest, the social value of research, data protection, open science and data sharing (ENERI Classroom n.d.), which is particularly relevant in the context of a pandemic.

Further, a direct relationship between the two is implied by the tendency to use "research ethics" as an umbrella term. For instance, in Norway "integrity" is included as one of the main features of the general principles of research ethics (Kaiser et al. 2022).

Most significant, however, is that research ethics and research integrity are inextricably linked via the values, intentions and actions of the person undertaking any research (the researcher). Researchers must accept that they act as moral (or immoral) agents. When researchers adopt the appropriate moral values/virtues, they strive to comply with requirements of both research ethics *and* research integrity (Chatfield and Law 2024).

While the credibility of a values-based code for research ethics has been demonstrated through the wide-scale adoption of the TRUST Code (TRUST 2018), this approach is new for a research integrity code. To understand why we believe the approach might also be effective for research integrity, it is helpful to consider why there were challenges to maintaining the integrity of research under the pressure of the COVID-19 pandemic.

Globally, the COVID-19 pandemic led to a surge in research output, with scientists rushing to publish findings about the virus, its transmission and potential treatments. The increase in volume, together with the accelerated pace, resulted in a flood of research publications, not all of which were of high quality. Concerns were soon raised about the reliability of findings and the potential for compromised peer review processes (Dinis-Oliveira 2020; Morens and Hammatt 2021; Lipworth et al. 2023; Evans et al. 2024). Fast-tracked research was associated with a decrease in rigor and quality, a rise in non-peer-reviewed publications and a competitive culture (Smith et al. 2023).

The increased use of open platforms for data sharing facilitated the rapid sharing of information, but also magnified risks. Data and research findings often bypassed traditional quality control measures, which enabled the spread of poor-quality studies across social media and other platforms. The rapid dissemination of flawed research not only harmed scientific credibility, but also contributed to misinformation (Dinis-Oliveira 2020; Evans et al. 2024).

The pandemic further exposed and exacerbated existing inequalities in the research ecosystem. For instance, questions arose about the continuation of non-COVID-19 studies and resource allocation (Lipworth et al. 2023; TENK 2024). Researchers working on non-COVID-19 projects faced limitations due to lockdowns and social distancing measures, which affected study completion and publication opportunities (Smith et al. 2023).

This imbalance perpetuated systemic inequities, both within research teams and globally. Teams and institutions in higher-income countries benefited disproportionately from well-established infrastructures, which enabled them to accelerate studies and share data more effectively than their counterparts in lower-income countries. Such disparities reinforced an already unequal playing field, raising concerns about equity and representation (Evans et al. 2024). While the pandemic initially fostered collaboration, travel restrictions and geopolitical tensions later hampered international research efforts (Smith et al. 2023).

During the COVID-19 pandemic, many researchers and research teams were placed under additional pressures for which they were not prepared and in facing which they were generally unsupported. Even within research teams, hierarchical disparities were accentuated as some already disadvantaged researchers bore heavier burdens than others (Inguaggiato et al. 2024). And pressures take their toll.

Numerous studies have indicated that the propensity to engage in research misconduct is associated with external pressures (Grimes et al. 2018; Houle et al. 2023). It is not that researchers do not understand that it is wrong to falsify data or exaggerate findings; it is rather that they tend to rationalise exceptions for themselves (Sticker 2017). Hence, the recommendation that an agent-based approach is needed to engender a greater sense of personal responsibility (Mitchell 2015). A values-based research integrity code is aligned with this thinking.

The PREPARED Code therefore has an ambitious aim in combining research ethics and research integrity in one code. It motivates researchers to act ethically by aligning the values between research ethics and research integrity to ensure moral clarity (Chatfield and Law 2024). By embracing the values of fairness, respect, care and honesty, researchers facilitate a culture in which both research ethics and research integrity can flourish.

5 Taking a Broad and Inclusive Approach to Code Development

In 2022, *Nature* adopted the TRUST Code to address helicopter research and ethics dumping across its entire science publishing portfolio (Nature 2022). In a podcast, Dr Sowmya Swaminathan, Nature's Director for Diversity, Equity and Inclusion, explained that the TRUST Code is

> a framework that's based on four values of fairness, respect, care, and honesty. It's a very comprehensive framework …. But at the same time, it's also designed in a way so as to make it relevant across multiple disciplines. So these are actually the elements that drew us to the code – the fact that they took such a broad, consultative approach, that they integrated the perspective of vulnerable populations, and that it is designed to be relevant across multiple disciplines (Kenneally 2022).

A very broad consultative approach and the integration of the perspectives of people in vulnerable situations were taken very seriously in the drafting of the TRUST Code, and also the PREPARED Code (see Chap. 5). This was important for the following reason. The PREPARED Code is built on 160 real-world risks in research ethics and research integrity as identified via the literature in nine languages. However, an obvious criticism of this approach is that the concerns of the least advantaged about matters involving research ethics and research integrity might not reach the academic literature. That is why broad consultations, including with the "Nairobi sex workers" and representatives from the South Africa San Council,[1] were key to building an inclusive ethics code. (For more information on the consultations, see Chap. 5.)

"Nairobi sex workers" is the term used by the PREPARED team for the more than 40,000 sex workers who are registered in ten foreign-funded research clinics in and around Nairobi that seek to prevent and treat HIV and other sexually transmitted diseases. Most of the sex workers have no income or support other than the meagre income derived from sex work. They live in small tin shacks, work well into middle age and accept dozens of clients every day because the payment from each is very low (Lucas et al. 2013). The COVID-19 pandemic seriously worsened the situation of this already highly marginalised group. In particular, increased poverty due to the loss of livelihoods brought about by COVID-19 restrictions led to heightened risk-taking behaviours, which in turn exacerbated their stigma (Schroeder et al. 2024).

The research foundation that represents the interests of the Nairobi sex workers in the PREPARED team (Partners for Health and Development in Africa, Kenya) has its own significant budget and was part of the team from the proposal-writing stage to the current dissemination phase.

The PREPARED team went one step further than the drafters of the TRUST Code, because pandemics impose extreme *additional* social and economic burdens, which the team also wanted to capture. Consequently, in addition to risk-identification from published literature and broad consultations, general challenges for the following (presumably disadvantaged) groups were scoped:

- sex workers in Nairobi informal settlements, the "Nairobi sex workers"
- health and social care workers in the United Kingdom and South Africa
- disabled people in the United Kingdom
- women researchers
- highly impoverished people in India.

The findings from these activities, summarised in Chap. 4, do not focus on research ethics and research integrity alone. A multitude of social and economic challenges were identified, including starvation during lockdowns (Kapoor 2020), the worsening of existing exclusions for disabled people (Partington et al. 2023) and the disproportionate burden on women, including women researchers (Inguaggiato et al. 2024), during the COVID-19 pandemic.

One might argue that such a broad view is unnecessary when writing a code primarily intended for use by researchers, research ethics committees and research integrity offices. However, the problems of the disadvantaged during the COVID-19 pandemic should not be ignored. Two critical lessons emerged from the PREPARED team's activities:

First, approaches to dealing with a pandemic in a high-income country must not be replicated uncritically in a lower-income setting. For instance, the COVID-19 lockdown the Kenyan government initiated, following the lead of governments in higher-income

---

[1] The government-recognised South African San Council was formed in 2001 to represent the interests of three major San communities, indigenous groups or First Peoples in South Africa. They were invited advisers at two PREPARED conferences and contributed to the PREPARED Code.

countries, had far-reaching consequences for the Nairobi sex workers, who were ill-equipped to deal with them.[2]

Second, the distribution of life-saving vaccines must be undertaken more equitably during future pandemics. A poignant example of inequity during the COVID-19 pandemic saw healthy adults in high-income countries being given booster doses of the vaccines while even medical staff in lower-income settings remained unprotected (COVAX 2022).

What impact did this very broad approach have on the PREPARED Code? Unlike the TRUST Code (TRUST 2028), the PREPARED Code is prefaced with a vision statement: "Pandemic research should be trustworthy and the results accessible to all." But there were also direct impacts on specific articles. Without the broad approach to ethics code development described above, Articles 6 and 8 might not have been included. Article 6 arises from the exacerbation of existing inequalities that occurred during the COVID-19 pandemic, especially for women researchers (Inguaggiato et al. 2024), something that should be avoided in future pandemics. Article 8 draws attention to the fate of community researchers, who can make a modest living by assisting researchers. During the COVID-19 pandemic, they were often classed as civilians, rather than researchers, and consequently lost access rights to research sites and related income during lockdowns. Article 8 of the PREPARED Code tries to preclude this from happening during the next pandemic. The direct impact upon the code shows that the broad approach was worth taking.

6 Short, Jargon-Free, Focused and Multidisciplinary

Codes for research ethics and research integrity come in all shapes and sizes. We are not suggesting that they should all look like the PREPARED Code. However, there are several reasons why the PREPARED team opted for a focused, short, jargon-free and multidisciplinary code.

First, it is important to remember that during pandemics, researchers who act unethically or without integrity may cause serious harm to research participants and others (Resnik 2024). Research ethics guidance specific to the challenges of the COVID-19 pandemic was hard to find, and it was needed in a hurry (Meagher et al. 2020). A short, focused and multidisciplinary code ensures that researchers from all fields can find – quickly and easily – guidance on aspects of responsible research specific to pandemics.

Secondly, keeping the code jargon-free helps reduce confusion and ensure that it is understandable across disciplines, and across a range of stakeholders. Ease of understanding is important, not just for researchers, research ethics committees and research integrity offices, the main target audiences of the PREPARED Code, but also for other users, such as research funders and sponsors, as well as research participants, NGOs and

---

[2] A short video produced by the PREPARED team in Nairobi can be seen at https://youtu.be/WwghcJr1F74. With 45,000 views at the time of writing (January 2025), this video is by far the most successful of the PREPARED project videos, indicating that it is beneficial when ethics projects do not stand back from ethics issues outside of their immediate focus.

the media. Brevity, simplicity, clarity and user-friendliness are highly valued in research ethics procedures:

> Whatever is brief and clear is better than what is not and saves time. What is simple and user-friendly is better than what is not even though the two have the same aims because it saves both time and mental energy. (Ouwe Missi Oukem-Boyer et al. 2016:1).

Additionally, a short and jargon-free ethics code can help research participants and NGOs resist unethical research, thereby providing a bottom-up tool to facilitate ethical research. For instance, the TRUST Code sparked the development of a sister code driven by an NGO, the South African San Council, which is now used in South Africa to stop exploitative research involving the Indigenous San (Schroeder et al. 2018 p. 21). Also, the PREPARED Code has already been translated into Swahili for use in the Nairobi clinics that are associated with the PREPARED project. The fact that the code is short and accessible enables its use with potential research participants.

Lastly, the PREPARED Code is multidisciplinary, though it is not intended to replace all other codes and guidelines for research ethics and research integrity. Most disciplines have their own tailored codes of research ethics: for instance, for physiologists who use animal experimentation, or for psychologists who conduct research with people who have mental health problems. These codes continue to apply during pandemics. Rather, the PREPARED Code is intended to *complement* other relevant codes of ethics and integrity. It provides guidance on issues that are exacerbated during pandemics.

To this end, the PREPARED Code does not replicate what already appears in the TRUST Code or the European Code of Conduct for Research Integrity, even though the risks of inequitable research and breaches of research integrity are likely to increase during a pandemic. Instead, these two codes are mentioned in the preamble to the PREPARED Code. Likewise for the Declaration of Helsinki, the primary source of ethics guidance globally for medical research involving humans (WMA 2024b), which itself might be regarded as brief and focused guidance.

For these reasons, and given the widespread and swift adoption of the TRUST Code, which is recognised for its brevity and user-friendliness, we hope that the PREPARED Code provides significant added value amid an abundance of existing ethics guidance.

7 Conclusion

For an ethics code to be effective, users must have confidence in its purpose and trust that it serves the interests of all stakeholders. But why should a new code enjoy trust? Given that there is no body of evidence to prove its worth, the answer to this question lies in the behind-the-scenes work of code development. The process of code development confers credibility (Messikomer and Cirka 2010), and knowing how a code was developed is a prerequisite to measuring its effectiveness (Kaptein and Schwartz 2008). Consequently, in this chapter we have sought to explain the guiding rationale for the development of the PREPARED Code and the inbuilt measures to help maximise its effectiveness and credibility.

The PREPARED Code is different from most other codes in a number of respects: it is risk-based, values-driven, multidisciplinary, short and jargon-free, it was developed in a highly inclusive process, and it is applicable to both research ethics and research integrity (Fig. 3). These factors enhance both the effectiveness and the credibility of the code.

Fig. 3. Measures to maximise the effectiveness and credibility of the PREPARED Code

The risk-based approach ensures the practicality and relevance of the PREPARED Code because it addresses real-world challenges that have occurred during pandemics. The alignment of the code with core moral values that resonate among people globally means that the code is more than simply a box-ticking exercise; it also serves to motivate and inspire users. The fact that the code is multidisciplinary, short and jargon-free means that when needed, research ethics and research integrity guidance is easy to find, and readily accessible to researchers and non-researchers alike. Finally, broad inclusivity is essential, not only for credibility (Messikomer and Cirka 2010), but also to ensure that the voices of the most disadvantaged are represented meaningfully.

References

Annas, G.J., Grodin, M.A.: The Nuremberg code. In: Emanuel, E.J., Grady, C.C., Crouch, R.A., Lie, R.K., Miller, F.G., Wendler, D.D. (eds.) The Oxford Textbook of Clinical Research Ethics, pp. 136–140. Oxford University Press, Oxford (2008)

Aristotle: The Nicomachean Ethics (Trans: Ross WD, Brown L). Oxford University Press, New York (2009)

Banks, S.: Cultivating researcher integrity: virtue-based approaches to research ethics. In: Emmerich, N. (ed.) Virtue Ethics in the Conduct and Governance of Social Science Research, pp. 21–44. Emerald Publishing Limited, Bingley (2018)

Black, J., Baldwin, R.: Really responsive risk-based regulation. Law Policy **32**(2), 181–213 (2010). https://doi.org/10.1111/j.1467-9930.2010.00318.x

Chatfield, K., Law, E.: 'I should do what?' Addressing research misconduct through values alignment. Res. Ethics **20**(2), 251–271 (2024). https://doi.org/10.1177/17470161231224481

Committee on Science, Engineering, and Public Policy: On Being a Scientist: A Guide to Responsible Conduct in Research, 3rd edn. The National Academies Press, Washington DC (2009). https://www.ncbi.nlm.nih.gov/books/NBK214568/pdf/Bookshelf_NBK214568.pdf. Accessed 6 Feb 2025

COVAX: COVAX calls for urgent action to close vaccine equity gap, 20 May 2022. World Health Organization (2022). https://www.who.int/news/item/20-05-2022-covax-calls-for-urgent-action-to-close-vaccine-equity-gap

Dawson, A.J.: Professional codes of practice and ethical conduct. J. Appl. Philos. **11**(2), 145–153 (1994). https://doi.org/10.1111/j.1468-5930.1994.tb00104.x

Dawson, A.J., Yentis, S.M.: Contesting the science/ethics distinction in the review of clinical research. J. Med. Ethics **33**(3), 165–167 (2007). https://doi.org/10.1136/jme.2006.016071

Dhai, A.: The research ethics evolution: from Nuremberg to Helsinki. S. Afr. Med. J. **104**(3), 178–180 (2014). https://doi.org/10.7196/samj.7864

Dinis-Oliveira, R.J.: COVID-19 research: pandemic versus "paperdemic", integrity, values and risks of the "speed science." Forensic Sci. Res. **5**(2), 174–187 (2020). https://doi.org/10.1080/20961790.2020.1767754

Dove, E.: Editorial: researching research integrity – and saying goodbye. Res. Ethics **20**(2), 137–142 (2024). https://doi.org/10.1177/17470161241235797

ENERI Classroom: Overlapping issues (n.d.). https://classroom.eneri.eu/overlapping-issues. Accessed 9 Sept 2024

Evans, N., Mlotshwa, L., Singh, M., et al.: Preliminary report on the expert validation workshop series (2024). (Unpublished report)

Franzen, M.: Science between trust and control: non-reproducibility in scholarly publishing. In: Atmanspacher, H., Maasen, S. (eds.) Reproducibility: Principles, Problems, Practices, and Prospects, pp. 467–485. Wiley, Hoboken (2016). https://doi.org/10.1002/9781118865064.ch22

Graham, D.J.: Why governments need guidelines for risk assessment and management. In: OECD Risk and Regulatory Policy: Improving the Governance of Risk, pp. 237–247. OECD Publishing, Paris (2010). https://doi.org/10.1787/9789264082939-en

Grimes, D.R., Bauch, C.T., Ioannidis, J.P.: Modelling science trustworthiness under publish or perish pressure. R. Soc. Open Sci. **5**(1), 171511 (2018). https://doi.org/10.1098/rsos.171511

Houle, F.A., Kirby, K.P., Marder, M.P.: Ethics in physics: the need for culture change. Phys. Today **76**(1), 28–35 (2023). https://doi.org/10.1063/PT.3.5156

Huttunen, S., Ojanen, M., Ott, A., Saarikoski, H.: What about citizens? A literature review of citizen engagement in sustainability transitions research. Energy Res. Soc. Sci. **91**, 102714 (2022). https://doi.org/10.1016/j.erss.2022.102714

Imperial College London: What is research misconduct? Research Office, Imperial College London (n.d.). https://www.imperial.ac.uk/research-and-innovation/research-office/research-governance-and-integrity/research-integrity/what-is-research-integrity/what-is-research-misconduct/. Accessed 6 Feb 2025

Ince, D.: The Duke university scandal: what can be done? Significance **8**(3), 113–115 (2011). https://doi.org/10.1111/j.1740-9713.2011.00505.x

Inguaggiato, G., Pallise Perello, C., Verdonk, P., et al.: The experience of women researchers during the Covid-19 pandemic: a scoping review. Res. Ethics **20**(4), 780–811 (2024). https://doi.org/10.1177/17470161241231268

Johnsson, L., Eriksson, S., Helgesson, G., Hansson, M.G.: Making researchers moral: why trustworthiness requires more than ethics guidelines and review. Res. Ethics **10**(1), 29–46 (2014). https://doi.org/10.1177/1747016113504778

Kaiser, M., Drivdal, L., Hjellbrekke, J., et al.: Questionable research practices and misconduct among Norwegian researchers. Sci. Eng. Ethics **28**, 2 (2022). https://doi.org/10.1007/s11948-021-00351-4

Kapoor, S.: Stories of hunger: India's lockdown is hitting the poorest. COVID-19 on South Asia Coronabrief, 21 April 2020. Friedrich Ebert Stiftung (2020). https://asia.fes.de/news/stories-of-hunger-indias-lockdown-is-hitting-the-poorest.html. Accessed 4 Feb 2025

Kaptein, M., Schwartz, M.S.: The effectiveness of business codes: a critical examination of existing studies and the development of an integrated research model. J. Bus. Ethics **77**, 111–127 (2008). https://doi.org/10.1007/s10551-006-9305-0

Kenneally, C.: Interview with Sowmya Swaminathan. Transcript: a global code of conduct for researchers. Velocity of Content, Copyright Clearance Center (2022). https://velocityofcontentpodcast.com/transcripts/a-global-code-of-conduct-for-researchers/. Accessed 5 Feb 2025

Khodadadyan, A., Mythen, G., et al.: Grasping the nettle? Considering the contemporary challenges of risk assessment. J. Risk Res. **24**(12), 1605–1618 (2021). https://doi.org/10.1080/13669877.2021.1894472

Kolstoe, S.E., Pugh, J.: The trinity of good research: distinguishing between research integrity, ethics, and governance. Account. Res. **31**(8), 1222–1241 (2023). https://doi.org/10.1080/08989621.2023.2239712

Kurzrock, R., Kantarjian, H., Stewart, D.J.: A cancer trial scandal and its regulatory backlash. Nat. Biotechnol. **32**(1), 27–31 (2014). https://doi.org/10.1038/nbt.2792

Law, E., Smith, I.: Ethical and informative trials: how the COVID-19 experience can help to improve clinical trial design. Res. Ethics **20**(4), 764–779 (2024). https://doi.org/10.1177/17470161241261768

Levine, C.: Has AIDS changed the ethics of human subjects research? Law Med. Health Care **16**(3–4), 167–173 (1988). https://doi.org/10.1111/j.1748-720x.1988.tb01942.x

Lipworth, W., Kerridge, I., Stewart, C., et al.: The fragility of scientific rigour and integrity in "sped up science": research misconduct, bias, and hype and in the COVID-19 pandemic. Bioeth. Inq. **20**, 607–616 (2023). https://doi.org/10.1007/s11673-023-10289-w

Lucas, J.C., et al.: Donating human samples: who benefits? Cases from Iceland, Kenya and Indonesia. In: Schroeder, D., Cook Lucas, J. (eds.) Benefit Sharing, pp. 95–127. Springer, Dordrecht (2013). https://doi.org/10.1007/978-94-007-6205-3_5

Meagher, K.M., Cummins, N.W., Bharucha, A.E., et al.: COVID-19 ethics and research. Mayo Clin. Proc. **95**(6), 1119–1123 (2020). https://doi.org/10.1016/j.mayocp.2020.04.019

Messikomer, C.M., Cirka, C.C.: Constructing a code of ethics: an experiential case of a national professional organization. J. Bus. Ethics **95**(1), 55–71 (2010). https://psycnet.apa.org/doi/10.1007/s10551-009-0347-y

Mitchell, L.A.: Integrity and virtue: the forming of good character. Linacre Q. **82**(2), 149–169 (2015). https://doi.org/10.1179/2050854915Y.0000000001

Morens, D., Hammatt, Z.: The COVID-19 pandemic: some thoughts on integrity in research and communication. Forensic Sci. Res. **6**(4), 310–315 (2021). https://doi.org/10.1080/20961790.2021.1980953

Nature: Nature addresses helicopter research and ethics dumping (editorial). Nature **606**(7) (2022). https://doi.org/10.1038/d41586-022-01423-6

Ogletree, T.W.: Value and valuation. In: Post, S.G. (ed.) Encyclopedia of Bioethics, vol. 5, 3rd edn., pp. 2539–2545. Macmillan Reference USA, New York (2004)

Ouwe Missi Oukem-Boyer, O., Munung, N.S., Tangwa, G.B.: Small is beautiful: demystifying and simplifying standard operating procedures: a model from the ethics review and consultancy committee of the Cameroon bioethics initiative. BMC Med. Ethics **17**(27), 1 (2016). https://doi.org/10.1186/s12910-016-0110-8

Partington, H., Garner, J., Chatfield, K.: Pushed closer to the edge: how the COVID-19 pandemic contributed to the increasing marginalisation of disabled people and their carers in the UK. A report for PREPARED (2023). https://prepared-project.eu/wp-content/uploads/2024/11/PREPARED-Experiences-of-disabled-people-in-the-pandemic-Final-.pdf. Accessed 4 Feb 2025

Paleco, C., García Peter, S., Salas Seoane, N., Kaufmann, J., Argyri, P.: Inclusiveness and diversity in citizen science. In: Vohland, K., et al. (eds.) The Science of Citizen Science, pp. 261–281. Springer, Cham (2021). https://doi.org/10.1007/978-3-030-58278-4_14

Pennock, R.T., O'Rourke, M.: Developing a scientific virtue-based approach to science ethics training. Sci. Eng. Ethics **23**, 243–262 (2017). https://doi.org/10.1007/s11948-016-9757-2

Resneck, J.S.: Revisions to the Declaration of Helsinki on its 60th anniversary: a modernized set of ethical principles to promote and ensure respect for participants in a rapidly innovating medical research ecosystem. JAMA **333**(1), 15–17 (2024). https://doi.org/10.1001/jama.2024.21902

Resnik, D.B.: What is ethics in research & why is it important? National Institute of Environmental Health Sciences (2024). https://www.niehs.nih.gov/research/resources/bioethics/whatis/index.cfm. Accessed 5 Feb 2025

Reverby, S.M.: Ethical failures and history lessons: the US public health service research studies in Tuskegee and Guatemala. Public Health Rev. **34**, 13 (2012). https://doi.org/10.1007/BF03391665

Rothstein, H., Borraz, O., Huber, M.: Risk and the limits of governance. Regul. Gov. **7**(2), 215–235 (2013). https://doi.org/10.1111/j.1748-5991.2012.01153.x

Schroeder, D., Chatfield, K.: Four universal values for a fresh new ethics framework and a global code of ethics. Royal Society for Public Health (2018). https://www.rsph.org.uk/about-us/news/guest-blog-four-universal-values-for-a-fresh-new-ethical-framework-and-a-global-code-of-ethics.html. Accessed 5 Feb 2025

Schroeder, D., Chatfield, K., Chennells, R., et al.: Vulnerability Revisited: Leaving No One Behind in Research. Springer, Cham (2024). https://doi.org/10.1007/978-3-031-57896-0

Schroeder, D., Cook, J., Hirsch, F., et al. (eds.): Ethics Dumping: Case Studies from North-South Research Collaborations. Springer, Cham (2018). https://doi.org/10.1007/978-3-319-64731-9

Schwartz, S.H.: An overview of the Schwartz theory of basic values. Online Read. Psychol. Cult. **2**(1) (2012). https://doi.org/10.9707/2307-0919.1116

Smith, E., Rakestraw, C., Farroni, J.: Research integrity during the COVID-19 pandemic: perspectives of health science researchers at an academic health science center. Account. Res. **30**(7), 471–492 (2023). https://doi.org/10.1080/08989621.2022.2029704

Sticker, M.: When the reflective watch-dog barks: conscience and self-deception in Kant. J. Value Inq. **51**, 85–104 (2017). https://doi.org/10.1007/s10790-016-9559-4

TENK: Research integrity in the time of COVID-19: Finnish research integrity barometer 2023. Finnish National Board on Research Integrity, Helsinki (2024). https://tenk.fi/sites/default/files/2024-11/TENK-Finnish-Research-Integrity-Barometer-2023-EN.pdf. Accessed 4 Feb 2025

Tiedemann, P.: Was ist Menschenwürde? Wissenschaftliche Buchgesellschaft, Darmstadt (2006)

TRUST: The TRUST code: a global code of conduct for equitable research partnerships (2018). https://doi.org/10.48508/GCC/2018.05. Accessed 5 Feb 2025

UKRIO: What is research integrity? UK Research Integrity Office (2023). https://ukrio.org/research-integrity/what-is-research-integrity/. Accessed 5 Feb 2025

UN: Universal declaration of human rights. A/RES/217(III). United Nations (1948). https://www.un.org/en/about-us/universal-declaration-of-human-rights. Accessed 5 Feb 2025

Varelius, J.: The value of autonomy in medical ethics. Med. Health Care Philos. **9**(3), 377–388 (2006). https://doi.org/10.1007/s11019-006-9000-z

Wilson, P.F.: Academic fraud: solving the crisis in modern academia. Exch. Interdiscip. Res. J. **7**(3), 14–44 (2020). https://doi.org/10.31273/eirj.v7i3.546

WMA: Background information on the Declaration of Helsinki. World Medical Association (2024a). https://www.wma.net/news-post/background-information-on-the-declaration-of-helsinki/. Accessed 5 Feb 2025

WMA: WMA Declaration of Helsinki: ethical principles for medical research involving human participants. World Medical Association (2024b). https://www.wma.net/policies-post/wma-declaration-of-helsinki/. Accessed 5 Feb 2025

Open Access This chapter is licensed under the terms of the Creative Commons Attribution 4.0 International License (http://creativecommons.org/licenses/by/4.0/), which permits use, sharing, adaptation, distribution and reproduction in any medium or format, as long as you give appropriate credit to the original author(s) and the source, provide a link to the Creative Commons license and indicate if changes were made.

The images or other third party material in this chapter are included in the chapter's Creative Commons license, unless indicated otherwise in a credit line to the material. If material is not included in the chapter's Creative Commons license and your intended use is not permitted by statutory regulation or exceeds the permitted use, you will need to obtain permission directly from the copyright holder.

Research Ethics and Integrity Challenges During Pandemics: The Research Foundation of the PREPARED Code

Pamela Andanda[1(✉)], Langelihle Mlotshwa[1], Orla Drummond[2], Vilma Lukaševičienė[3], Giulia Inguaggiato[4], and Klaus Leisinger[5]

[1] School of Law, University of the Witwatersrand, Johannesburg, South Africa
pamela.andanda@wits.ac.za
[2] Trilateral Research, Belview Port, Ireland
[3] The Division of Medical History and Ethics in the Medical Faculty of Vilnius University, Vilnius, Lithuania
[4] Department of Ethics, Law and Humanities, Amsterdam Public Health Institute, Amsterdam UMC, Vrije Universiteit Amsterdam, Amsterdam, The Netherlands
[5] Foundation Global Values Alliance, Basel, Switzerland

Abstract. Crises and public health emergencies can have devastating and wide-ranging impacts across healthcare, social, cultural, economic and political contexts. Against this backdrop, many research ethics and research integrity challenges can be exacerbated, and new ones can emerge as researchers and other stakeholders find themselves in testing and unstable environments. Learning from these challenges, to prepare for similar future crises, requires a broad perspective and encompassing vision. It also requires careful identification and analysis of challenges to ensure that guidance for future crises has real-world applicability. A risk-based approach to ethics code development begins with the identification of ethics risks or challenges, which itself requires extensive research. This chapter describes the research foundation upon which the PREPARED Code was built. It presents an overview of the results of an in-depth review of published literature in nine languages, additional scoping reviews on Ebola and avian flu, investigations into the challenges experienced by groups who were disadvantaged during the COVID-19 pandemic, an analysis of the human rights challenges in the context of sudden global crises, an investigation into how one pharmaceutical company overcame governance challenges to produce a vaccine in record time, and an analysis of pandemic research ethics guidance documents. Together the findings from these activities constituted a strong foundation for the development of the PREPARED Code.

Keywords: Research ethics · research integrity · pandemic ethics

1 Introduction

There are various ways to build the foundation for new ethics codes. As outlined in Chap. 3, the PREPARED team chose the risk-based approach combined with extensive consultations. This required identifying and analysing emergent and exacerbated research ethics and integrity challenges during pandemics.

In this chapter, we describe the main research results that provided the foundation for the PREPARED Code. In particular, we present selected findings from:

- a review of the published literature from the COVID-19 pandemic in nine languages (English, Chinese, French, German, Hindi, Japanese, Korean, Russian and Spanish) and additional scoping reviews in English on Ebola and avian flu epidemics (Sect. 2)
- investigations into the challenges experienced by groups who were probably disadvantaged because of the pandemic, including health and social care workers, people with disabilities, women researchers and highly marginalised groups such as the Nairobi sex workers (Sect. 3)
- an analysis of human rights challenges in the context of sudden global crises (Sect. 4)
- an investigation of how one pharmaceutical company overcame governance challenges to produce a vaccine in record time (Sect. 5)
- an analysis of pandemic research ethics guidance documents (Sect. 6).

The research ethics and integrity challenges identified through the above methods were validated in workshops with researchers, policymakers and advisers, research ethics and research integrity experts, and patient groups and their representatives, as described in Chap. 5.

2 Literature Reviews in Nine Languages

To capture as many challenges as possible for research ethics and research integrity during times of crisis, and to avoid linguistic bias, literature reviews were conducted in English, Chinese, French, German, Hindi, Japanese, Korean, Russian and Spanish. These reviews focused primarily upon the COVID-19 pandemic, but additional scoping reviews were undertaken in English on the Ebola epidemic, which the World Health Organization (WHO) declared a "public health emergency of international concern" in 2014 (Gostin et al. 2014), and on the avian flu epidemic (Mittal and Medhi 2007).

The types of literature searched included peer-reviewed and non-peer-reviewed academic literature, grey literature, articles in the media, and official guidance and advice, depending on where the best sources of information could be found in each language. This helped maximise the number of risks identified. Three of the reviews, Korean, German and Chinese, have been published in full elsewhere (Park and Kim 2024; Seedall and Tambornino 2024; Zhu et al. 2024).

Together, the reviews generated a vast amount of rich data that was subsequently pooled for analysis and led to the identification of 160 challenges for research ethics and research integrity (see Chap. 5). Since it would not be possible to describe all 160 identified challenges here, this section includes a synopsis of the primary challenges that were uncovered for various stakeholders in the research process. Examples from the different language reviews are provided for illustrative purposes.

2.1 Challenges for Research Ethics Committees

Research ethics committees have a vital role during pandemics. They must help to ensure that research is conducted ethically and safeguard the rights and wellbeing of participants, while facilitating rapid scientific advancements (Tamariz et al. 2021). Numerous challenges for research ethics committees during the COVID-19 pandemic were reported across many of the language reviews. For instance, the English review revealed how research ethics committees experienced increased workloads and demands for rapid review of research protocols due to the urgency brought about by the pandemic (Marzouk et al. 2021; Shekhani et al. 2021; Tamariz et al. 2021; Kadam et al. 2022).

Similarly, the Spanish literature describes the widespread sense of urgency during the pandemic that led to the acceleration of research pathways. This inevitably put pressure on the ethics review processes in Spanish-speaking regions (Barajas and Valderas 2020; Espinoza-Navarro and Rivera-Gutiérrez 2021; Mendoza and Abreu 2021). For instance, in Spain, where research ethics committee members are normally also healthcare professionals, the frequency of the review meetings increased from once a month to once a week, adding to their existing burdens (Bugarín-González et al. 2020).

In Korea, where research ethics approval normally takes one or two months, the government requested that the review process be shortened to less than a week (Park and Kim 2024). With additional workloads, time pressures and the switch to online working, research ethics committees were forced to seek alternative ways to function during the pandemic, balancing the acceleration of review processes with the maintenance of methodological rigour (Barajas and Valderas 2020; Mendoza and Abreu 2021).

The English review also highlighted challenges associated with a lack of precise guidelines, especially in the early stages of the pandemic (Marzouk et al. 2021). New trial designs presented challenges, with regular adjustments of studies to reflect evolving guidelines on COVID-19 (Marzouk et al. 2021). The protocols for these new designs required general expertise in ethics and public health preparedness, which was not always available due to the demands on members to review more proposals within a limited time (Tamariz et al. 2021). Some protocols were of poor quality as they were hastily prepared by eager researchers who wanted to join the race for research results (Marzouk et al. 2021; Shekhani et al. 2021), thus compelling committee members to spend more time on the scientific quality of the protocols than on the ethics (Shekhani et al. 2021). Additionally, and in spite of their extra efforts, the opinions of research ethics committees were not always respected. For instance, out of 42 sites in a multicentric trial in India, the decision of the Central Ethics Committee was followed by only three sites (Bassi et al. 2022).

2.2 Challenges for Participants

During the COVID-19 pandemic, the urgent need to develop treatments and vaccines often led to expedited research processes, compromising the thoroughness of informed consent (Goldman and Gelinas 2021). For instance, the Chinese review reported that after the outbreak of the pandemic, there were three main concerns about informed consent (Cheng et al. 2020; Ding et al. 2020; Zhang 2020; Hu and Dong 2021): first, that signing informed consent forms in the isolation wards might cause contamination; second, that signing these forms would be difficult for potential human participants who were critically ill and confined to isolation wards with no family members or legal representatives; and third, that the desire to obtain informed consent speedily for urgent projects might create a type of "therapeutic misunderstanding" on the part of patients, through insufficient explanation by researchers and the patients' own eagerness to get treatment. This could result in neglect of the risks associated with participation in the research.

In Korea, because it was difficult to obtain consent through face-to-face discussions with patients who were quarantined and undergoing treatment for COVID-19, several alternatives were proposed, including telephone or video explanations, consent from legal representatives, and the storing of photographs of consent forms as a substitute for written consent (Shin 2020). And ethical breaches of the requirements for informed consent in clinical research did occur. For instance, the French press reported breaches of informed consent requirements in at least three research projects in one institution alone (Larousserie 2022).

Balancing the potential benefits of research against the risks to participants became more complicated during the pandemic. The high transmission rate of COVID-19 and limited knowledge about the virus meant that researchers were working in a climate of uncertainty, which made it difficult to be confident about participant safety (Bierer et al. 2020). The most extreme example of this challenge could be seen in the vigorous debate regarding the permissibility of human challenge studies during the early phases of the pandemic, before effective COVID-19 vaccines were developed.

Human challenge studies – that is, the intentional infection of healthy volunteers with the virus in a controlled environment – were widely proposed as a quick way of gaining insights into COVID-19, as well as of speedily testing potential treatment and vaccine candidates (Chappell and Singer 2020). However, concerns were raised about lowering ethical standards to enable these studies to proceed, with unknown risks to participants (Weijer 2024). In Korea, academic papers examined the risks, benefits and ethical considerations of participant consent, analysing historical cases of human challenge trials for diseases like dengue fever, cholera and Zika virus (Choi 2020; Fang et al. 2020; Jung and Kim 2020; Lee 2020). In the German literature, the intentional infection of participants stirred painful memories of the Nazi atrocities in Europe (Jamrozik and Selgelid 2020), but many researchers recommended human challenge studies for their scientific potential, and because there was a lack of comparable alternative methods to achieve rapid and accurate results (Faust et al. 2020). However, strict approval procedures requiring painstaking risk calculations discouraged German scientists from opting for such studies (Tambornino and Lanzerath 2020).

2.3 Challenges for Researchers

Researchers were also affected during the COVID-19 pandemic by demands related to the change in work patterns and the pressure to generate helpful, evidence-based information. The situation in Germany was not untypical: following the rise of #ichbinhanna (I am Hanna), a viral campaign evolved, amplifying the voices of researchers at German universities struggling with job insecurity and excessive workloads. The pandemic was widely regarded as the tipping point in academic workplace dynamics (Mittermeier 2021). It worsened the already precarious working conditions at many universities and research institutions. The COVID-19 crisis deepened existing inequalities in academia, particularly affecting women researchers. As remote work became the norm, or was even mandated, across much of German-speaking Europe, women were increasingly expected to manage childcare and household responsibilities, leaving less time for scientific work (Taschwer 2022). Consequently, women-authored publications declined, and their citation rates dropped in comparison to male authors, a trend shown to undermine research integrity (Miller, Valeva, Prieß-Buchheit 2022).

2.4 Challenges for Healthcare Staff

A vast body of literature has explored the challenges faced by healthcare workers during the COVID-19 pandemic (Ehrlich et al. 2020), with many also taking on research responsibilities. For example, in Wuhan, China, the city where COVID-19 was first identified, when the number of patients surged exponentially in the early stages of the outbreak, shortages of personnel forced healthcare providers to juggle both patient care and scientific research, and they struggled to balance these demanding roles (Wu et al. 2021). Additionally, the heavy workload often left them with insufficient time and energy for follow-up visits, limiting their ability to collect essential patient data needed for research completion (Liu et al. 2020).

2.5 Challenges for Societies

The primary challenges for societies that emerged during the COVID-19 pandemic were associated with the trustworthiness of science and the messages relayed via the media. For instance, the Spanish review revealed that the pandemic contributed to a lack of rigour in articles published as pre-prints, or fast-tracked for publication, that were later retracted (Dadalto et al. 2020; Bermúdez and Maldonado 2021). The haste with which articles were being published also increased the risk of irresponsible research practices such as plagiarism, duplicate publication, falsification, fabrication, gift authorship, conflicts of interest and inadequate peer review (Bermúdez and Maldonado 2021). And in India, Todhunter (2021) reported a lack of rigour in the interpretation of research findings leading to the dissemination of misleading information.

People were desperate for news of a cure. Early in the pandemic, the potential use of hydroxychloroquine as a treatment for COVID-19 became a talking point in France when a professor announced the "endgame" against the new coronavirus. He claimed that hydroxychloroquine, a synthetic derivative of quinine and normally prescribed for malaria, could inhibit the virus within a few days (jmichel2you 2020). The controversial use of this drug spread quickly around the world, as could be seen, for instance, in India (Bangalore et al. 2020; Chaturvedi et al. 2020) and Germany (Christian 2022), where the debate provided an impetus for the rigorous scrutiny of studies.

The Russian review reported how the speed with which the COVID-19 Sputnik vaccine was developed and approved raised concerns about its safety and efficacy (Bucci et al. 2020). Russia became the first country in the world to approve a COVID-19 vaccine for widespread public use in August 2020, but the vaccine's efficacy and safety were allegedly announced before the clinical trials had been completed or data published

(Cohen 2020). It was also alleged that vaccine trial data may have been manipulated and duplicated (Heidt 2022).

The lack of effective coordination of clinical studies had impacts on the quality of scientific data (Bompart 2020). As reported in the German literature, a lack of upstream coordination resulted in high overlap, high competition and, subsequently, a recruitment crisis that bred methodological weaknesses (Faust et al. 2020; Hirt et al. 2021; Pearson 2021), all of which had an impact on the reliability of study findings.

A further challenge for society concerned the routine exclusion of vulnerable individuals and groups (like pregnant women and elderly persons) from research. The practice may have prevented certain groups from accessing the benefits of research, ultimately rendering them more vulnerable. This challenge also generated debate around the world. For instance, in Korea it was argued that the exclusion of individuals or groups solely because of their vulnerability, without scientific or ethical justification, was not acceptable (Yoo and Kim 2021).

2.6 Global Challenges

The COVID-19 pandemic exacerbated existing socioeconomic vulnerabilities and inequalities globally. Much of the identified Spanish literature referred to global inequality in research risks, burdens, benefits and resource distribution. Some of it stressed the need to pay extra attention to potential cases of ethics dumping in times of crisis, highlighting that the rights and safety of vulnerable populations should not be overlooked for "the greater good" (Lopez 2021; Schveitzer and Thome 2021), or the risks and benefits of COVID-19 research unfairly allocated (Flores 2020; Manchola-Castillo 2022).

The issue of open data-sharing also brought inequalities and prejudice to light. The WHO advises that all parties who are involved in public health surveillance should share data in a timely fashion (WHO 2017, Guideline 15). Nevertheless, although there were calls for data-sharing in the spirit of promoting open science during the pandemic (Kadakia et al. 2021), Southern African countries were unjustly ostracised and subjected to travel bans by high-income countries when their scientists shared the genomic sequencing data of the omicron variant of COVID-19 (B.1.1.529) with the global community (Mallapaty 2021; Moodley et al. 2022).

2.7 Comparison with Ebola and Avian Flu

We do not know whether there will be another pandemic. If there is, we cannot be certain that the research ethics and research integrity challenges will be the same as those that arose during the COVID-19 pandemic. Consequently, the PREPARED team also searched for research ethics and research integrity challenges that were reported during the Ebola and avian flu epidemics.

The key differences between the COVID-19, Ebola and avian flu outbreaks lay in the nature of the infectious agents, the scale of the outbreaks and their geographical distribution. While COVID-19 had a global impact, affecting nearly every country, Ebola and avian flu remained more geographically contained. As a result, there were significantly fewer articles identified as relevant to the scoping reviews on Ebola and avian flu (ten and nine respectively). Nevertheless, common themes emerged for the three diseases, including the difficulties of balancing rapid research deployment and rigorous ethical standards, the challenges of conducting equitable research in resource-limited settings and multiple issues related to misinformation and rushed publication.

One area of difference, due to the nature of the virus, was that biosafety concerns were particularly prominent for avian flu. Ethical debates arose around gain-of-function research, which involves enhancing or introducing new functions through genetic manipulation (Swazo 2013). In the context of infectious disease research, gain-of-function may alter the pathogenicity, infectivity, transmissibility or host range of the pathogen. This raises concerns about dual use, as research aimed at understanding and mitigating threats could also be misused to develop bioweapons (Shinomiya et al. 2022).

Additionally, Ebola outbreaks underscored the importance of community trust, engagement and culturally sensitive research practices. Mistrust of foreign medical interventions led to resistance in some affected regions, reinforcing the need for ethical frameworks that respect local customs and actively involve community stakeholders in decision-making processes (Wilhelmy et al. 2022).

3 Challenges Experienced by Groups Who Faced Extraordinary Burdens During the COVID-19 Pandemic

Given the widespread and complex impacts of the COVID-19 pandemic, the PREPARED team recognised the importance of embedding research ethics procedures within a broader social framework. Drawing on Horton's (2020) concept of the COVID-19 pandemic as a *syndemic*, the team acknowledged that addressing underlying social and economic inequalities was essential for ensuring that nations were ethically prepared for future pandemics. Relying solely on biomedical questions, such as the search for treatments and vaccines, would not resolve broader health crises fully. Viewing the COVID-19 pandemic as a syndemic highlights its deep social roots and the interconnectedness of the virus with other socioeconomic factors that disproportionately affect disadvantaged individuals and groups (Horton 2020).

At the heart of this approach is a dedication to social justice, achieved by elevating the voices and experiences of key populations who faced disproportionate hardships during the COVID-19 pandemic.

Accordingly, the PREPARED team learnt from the experiences of

1. people on the poverty line who faced severe additional threats to their livelihoods
2. people with disabilities
3. groups that suffered disproportionate burdens due to their gender
4. frontline personnel in the health and social care sector.

By integrating these diverse perspectives, the PREPARED team aimed to ensure that future research endeavours would be both ethically sound and socially responsive, addressing the complex realities of those most affected by global crises. In the following paragraphs, we provide a short summary of the key challenges experienced by the groups we investigated.

3.1 People on the Poverty Line Who Faced Severe Additional Threats to Their Livelihoods

Thanks to their long-standing community engagement programmes, Partners for Health and Development in Africa (PHDA) were well positioned to organise a bottom-up consultation with sex workers and healthcare providers in Nairobi to inform the PREPARED Code. The challenges encountered between March and December 2020, during the COVID-19 pandemic, by those enrolled for HIV prevention and treatment services at ten clinics run by PHDA in the Nairobi area (Kimani and Adhiambo Odhiambo 2023) were discussed.

This consultation enabled the PREPARED team to learn from a highly marginalised group that was severely affected by COVID-19. The COVID-19 pandemic had a devastating impact on sex workers, exacerbating stigmatisation and discrimination and exposing them to increased violence from clients, police and society. Abrupt restrictions like curfews and lockdowns threatened their livelihoods, forcing many into risky behaviours such as unprotected sex, due to financial constraints and the closure of clinics providing essential supplies. These hardships were combined with stigmatising COVID-19 testing and treatment practices, as well as restrictions on funerals and weddings, which deprived communities of traditional grieving processes and sometimes led to post-traumatic stress disorder.

Having to work at home exposed sex workers to discrimination from neighbours and heightened stigma, while the shift to social media to find new clients increased the risks of violence: sex workers reported that the risk of rape or that of unprotected forced sex was higher with clients met through social media than those encountered face to face or via existing contacts. These tribulations, coupled with isolation from social support networks, posed a severe threat to mental health. Additionally, a lack of understanding about COVID-19 safety measures deepened mistrust and fear. Combined, these factors increased stigma and poverty and led to higher risks of HIV infection (Schroeder et al. 2024). The overlapping challenges aggravated the systemic vulnerabilities and social marginalisation sex workers faced during the pandemic (Kimani and Adhiambo Odhiambo 2023).

3.2 People Living with Disabilities

During global crises like the recent COVID-19 pandemic, people with disabilities are in an especially vulnerable situation (Shakespeare et al. 2021). The pandemic served to exacerbate their potential marginalisation through service disruptions, public health measures and their greater risks of adverse outcomes from the virus (Partington and Chatfield 2023b).

Consultations with disabled people and their carers were conducted in collaboration with Comensus, a service-user and carer community group that enables people with disabilities and their carers to participate in research and training. Thirteen participants contributed through informal conversations, either individually or in groups, or by submitting reflective thoughts in writing or via audio recordings. Participants reported that the measures taken to mitigate the pandemic led to increased marginalisation, social injustices and failure to uphold the rights of people with disabilities and their carers. The main challenges reported by these groups related to disrupted and deteriorating services, the fact that life got and stayed smaller because of lockdowns and restrictive measures, and the difficulties of grappling with new rules and new information (Partington and Chatfield 2023b).

3.3 Disproportionate Burdens Due to Gender

The experiences of women researchers were explored through a scoping review of qualitative literature on their experiences during the pandemic (Inguaggiato et al. 2024), as well as qualitative data collection and analysis conducted in Cyprus through conversations with women researchers from various fields (Antoniou et al. 2024). These studies revealed common experiences among women researchers worldwide. Researchers reported an increase in care responsibilities both at home and at work, a decrease in academic productivity impacting career growth, a lack of support from institutions and family members, and difficulty in reconciling and managing conflicts between private and professional identities and roles. Additionally, women researchers faced forced flexibility, challenges adapting to new research methodologies, and difficulties with online teaching mandated by lockdowns. Overall, the pandemic exacerbated existing gender inequalities in the research professions and reinforced gendered power dynamics in academia.

3.4 Frontline Personnel in the Health and Social Care Sector in the UK and South Africa

Understanding the challenges healthcare workers face is crucial in contextualising research conducted during pandemics. Since healthcare workers are directly or indirectly affected by pandemic-related clinical studies, their perspectives must be included in the creation of ethics guidelines for conducting research during pandemics (Partington and Chatfield 2023a).

The PREPARED team studied the experiences of frontline personnel in the UK (Partington and Chatfield 2023a), and social and healthcare workers in South Africa (Mlotshwa et al. 2023). These two countries recorded some of the highest numbers of COVID-19 cases and deaths in Europe and Africa, respectively (Mbunge 2020; Konstantinoudis et al. 2022).

To gather the experiences of social and healthcare workers in South Africa, the team undertook a scoping review of relevant literature. For the frontline personnel in the UK, a meta-ethnographic analysis of published qualitative studies was conducted. Common findings from the two approaches revealed that professionals working in the health and social care sectors experienced difficulties in adapting to sudden changes, expressed a need for support and leadership, and reported physical and emotional burdens as well as safety concerns (Mlotshwa et al. 2023; Partington and Chatfield 2023a).

4 Analysis of Human Rights Challenges in the Context of Sudden Global Crises

To explore the underpinning legal issues relating to crises and emergencies, the PRE-PARED team carried out a legal and human rights analysis. This included a thorough scoping of the main legal sources, human rights instruments and reports from committees and special rapporteurs. This information was coded via a thematic framework informed by key legal and human rights issues in times of crisis. Following the creation of this framework, an extensive literature review was conducted for the emergent themes, with each theme subjected to a dedicated literature search.

This analysis also aimed to highlight the wider challenges of the global COVID-19 pandemic, assuming that understanding of the vulnerabilities endured by marginalised communities during crises and emergencies leads to enhanced ethical research practice (Drummond 2023). Findings were themed into three main sections:

1. legal, structural and overarching issues
2. human rights obligations
3. marginalised communities.

4.1 Legal, Structural and Overarching Issues

The analysis centred on the rule of law during crises and emergencies, the development of administrative laws during these times, and the use of emergency laws and powers. A key finding was that government responses to COVID-19 provoked a range of national legal responses aimed at galvanising the mitigation of negative impacts. However, these legal mechanisms had far-reaching effects on fundamental rights and freedoms as the legal landscape shifted precipitately to block the transmission of the virus. The rapid implementation of legislative and regulatory solutions raised several concerns, particularly in relation to the erosion of elementary foundations of the rule of law. These included, but were not limited to, ad hoc lawmaking decisions, the imperfect drafting quality of legislation, the lack of consultation processes or scrutiny, and the accelerated promulgation of laws.

Admittedly, crises and emergencies pose significant challenges to governing structures, but this does not lessen the responsibility of those structures to ensure that the main tenets of the rule of law are adhered to (Cormacain 2020). The COVID-19 pandemic generated a significant volume of primary and secondary legislation over a very short period, but, crucially, the rule of law still requires proper accountability and lawmaking procedures to be followed, even in times of emergency and crisis (Cormacain 2020).

To prepare for any future crisis or emergency, it is essential that governments learn lessons from the legal and regulatory responses to COVID-19, acknowledge the weaknesses of those responses, and develop frameworks for future solutions. Any plans for legal preparedness must ensure that the rule of law forms the elemental basis of all future action and aids the protection of fundamental rights and freedoms.

4.2 Human Rights Obligations

The exploration of human rights obligations identified key strands in relation to the COVID-19 pandemic. These included issues such as the right to health and healthcare, and the protection of healthcare professionals and frontline workers. In addition, the analysis explored vaccine development from a human rights perspective, including equitable global access and the overarching right to science. Taking into account the rise of disinformation and misinformation, the analysis included the right to accurate information and the right to the protection of privacy. Also addressed were human rights in the context of global food security and access to housing, water, sanitation and hygiene, and the protection of the environment during times of crisis and emergencies.

Being prepared for future crises and emergencies means that effective international cooperation and solidarity must be ensured to safeguard the protection of human rights. To enable this, structural inequalities in relation to preparedness have to be corrected. The analysis acknowledged that infrastructural underinvestment in valuable support services was one of the key areas for examination and mitigation. In essence, to prepare for any future healthcare crisis, global inequities need to be addressed head-on. Investment in healthcare systems is urgently required. Additionally, further strengthening of labour laws is needed to protect frontline workers. Vaccine inequity must be eradicated. There is a necessity and an obligation upon states to ensure access to accurate information and to protect personal data during crisis and emergency.

In addition, the analysis noted that while it was vital to adopt a human rights approach to emergency and crisis responses, it was even more imperative to eradicate marginalisation in normal times. Human rights violations such as homelessness, disability, inequality, lack of access to essential healthcare, lack of access to water and sanitation, and gender-based violence must be eradicated. The analysis concluded that these were the challenges faced today that needed to be tackled to be better prepared for tomorrow.

4.3 Marginalised Communities

The human rights analysis noted that crises and emergencies could compound existing inequalities and worsen human rights violations, or create new ones, for many communities. During the development of the thematic framework, several distinct communities that were detrimentally impacted by the COVID-19 pandemic were identified. These included:

- poorer communities
- children and young people
- women and girls
- LGBTQ+
- minorities
- indigenous people
- migrants, displaced people and refugees
- older people
- people in detention or in institutions
- persons with disabilities

While the needs of specific groups were considered, it was also acknowledged that individuals might well belong to more than one of these groupings, thus layering inequalities into entrenched and intractable patterns. Although they were separated into distinct themes for the purposes of the analysis, members of each identified community are not homogeneous. These communities do not exist in silos of inequality. Overlapping and intersectional inequalities therefore must be taken into account when considering individual needs, human rights and legal obligations.

Ultimately, "[i]n order to prepare for the next crisis or emergency, it is essential that no one is left behind. There is a need for a global commitment to responding or preparing for crises and emergencies in a way that is sensitive to the most marginalised communities" (Drummond 2023). Strategically, it is essential that the communities identified above be included at the planning stages of crisis preparedness and that their experiences of crisis and crisis management be taken into account. As the meaningful participation of all citizens is an assured human right, this approach is essential to provide an empowering environment where marginalised communities can retell their experiences and help shape future planning.

5 Overcoming Governance Challenges in the Pharmaceutical Industry

In addition to looking at the experience of marginalised populations and the human rights situation during the COVID-19 pandemic, the PREPARED team also examined the successful story of BioNTech, which "won ... the race for a COVID-19 vaccine ... without purposefully infecting healthy participants with an infectious agent that can cause severe illness or death and for which no rescue therapy had existed" (Leisinger and Schroeder 2024: 847). The biotechnology company developed a life-saving vaccine with over 90% efficiency in less than one year and in accordance with existing principles of good clinical practice.

There are three key lessons to be drawn from the case.

First, the fact that the Paul Ehrlich Institute and BioNTech had been engaged in a professional dialogue for many years created an atmosphere of mutual trust. On this basis, and given the urgency of the situation, a presentation slot for a planned vaccine study was provided within a week rather than within three months (Leisinger and Schroeder 2024: 849).

Second, while all existing good clinical practice regulations were adhered to, increased efficiency was achieved by combining and overlapping different development phases and by regulators implementing a rolling review of clinical trial data.

Third, the development process was also accelerated by risking the security of company assets (BioNTech vaccine candidates) and partnering with the much larger pharmaceutical giant Pfizer, with only a letter of intent in place. This was most unusual, because a letter of intent does not protect assets that have been shared. The drafting process for a full collaboration agreement normally takes at least six months (Miller, Türeci, Şahin 2022).

During this time, no proprietary technology (like BioNTech's vaccine candidates) would normally be shared. One day after the letter of intent was signed, Uğur

Şahin, the co-founder of BioNTech, instructed his disbelieving team to "share everything". (Leisinger and Schroeder 2024: 853).

This case study showed that the speed of vaccine development can be accelerated significantly while research ethics and research integrity values are preserved.

6 Analysis of Pandemic Guidance Documents

Chapter 3 notes that the PREPARED team combined a risk-based approach to ethics code development with extensive consultations. At the same time, the team undertook a detailed gap analysis of existing ethics codes to check that no relevant ethics issues had been overlooked. In line with the reasonable assumption that code- and drafter-based guidance might not always link to real world challenges but may sometimes be based on conjecture (see Chap. 3), challenges that came from the gap analysis were then to be verified through a new search of the literature and consultation with experts. So, what did we find?

To identify relevant ethics codes for the analysis, the primary search for the PREPARED team was undertaken by the senior librarian, Kelly Laas, at the Illinois Institute of Technology in Chicago. This library hosts the largest collection of ethics codes in the world. The search, undertaken in 2023, was limited to documents in the English language and focused on ethics and integrity guidance relevant to global crises and pandemics. It generated 103 ethics documents on global crises and pandemics, with 36 documents having been issued prior to 2020 and almost twice as many (67) since 2020.

The ethics documents identified by the Chicago search were reviewed manually by the PREPARED team to identify further relevant documents based on their references. Additional searches were also undertaken in the following databases: Council of Europe Bioethics COVID-19, WHO publications, the OECD iLibrary, national competent authorities and Council of Europe national ethics committees. This process identified another 133 ethics documents.

Documents from the Chicago search and the additional search (236 codes, recommendations and guidelines) were reviewed manually and selected for further analysis if they met the following inclusion criteria: the document was adopted during the COVID-19 pandemic, at least part of the document was relevant to the conduct of research during the COVID-19 pandemic, it was written in English, and it was issued by an international or national institution, organisation or association. The search was not limited to a specific country or region. Ninety-seven documents met these inclusion criteria. Most had been adopted by national government agencies (42), followed by international organisations (28), ethics bodies (11), professional associations (7), nongovernmental organisations (3), scientific councils (3) and universities (3).

Content analysis was used to examine the content and contextual meaning of selected documents (Hsieh and Shannon 2005). Documents were analysed using MAXQDA Analytics Pro software (2022 version) for coding into the main predefined categories: ethics and integrity issues (challenges), virtues, principles and articles.

The challenges were extracted from the database and compared with the challenges derived from the primary research undertaken by the PREPARED team, that is, the reviews in nine languages. How the challenges were formulated in existing ethics guidance helped refine the PREPARED Code (e.g. see Chap. 5 on the addition of "quality-controlled" to the data-sharing article). However, none of the challenges extracted had not already been identified by the risk-based analysis.

7 Conclusion

This chapter provides an illustrative overview of the findings that constitute the research foundation for the PREPARED Code. Reviews of the published literature from the COVID-19 pandemic, additional scoping reviews on the Ebola and avian flu epidemics, empirical and literature-based studies revealing general challenges for groups in vulnerable situations, and findings from the human rights analysis, taken together, all ensure that the PREPARED Code is built on real-world risks. The analysis of pandemic and crisis guidance documents helped us confirm that the risk-based approach had not overlooked any major challenges.

The argument of Sutrop and colleagues that "the process of drafting codes of ethics should be as inclusive as possible" (2020: 81) can be applied to the development of the PREPARED Code. Not only was the research basis broad, including the perspectives of those in vulnerable situations, but findings were also subsequently validated through workshops and consultation with diverse stakeholders such as policymakers and advisers, experts from the pharmaceutical industry, research ethics and research integrity colleagues, and patient groups and their representatives (see Chap. 5).

By letting the views and experiences of marginalised groups take centre stage, we have endeavoured to ensure that the PREPARED Code will support those who need it the most. In times of crisis, no one should be left behind. Therefore, preparedness should be planned in a way that is sensitive to the most marginalised communities. This includes the empowerment of these communities to relay their experiences, their thoughts and concerns, and their opinions on the best courses of action for their needs.

References

Antoniou, J., Kornioti, N., Antoniou, K.: Navigating the post-pandemic normal: learning from the experiences of cyprus-based female researchers during the COVID-19 pandemic. Soc. Sci. **13**(6), 280 (2024). https://doi.org/10.3390/SOCSCI13060280

Bangalore, S., Sharma, A., Slotwiner, A., et al.: ST-segment elevation in patients with Covid-19: a case series. N. Engl. J. Med. **382**(25), 2478–2480 (2020). https://doi.org/10.1056/nejmc2009020

Barajas, A.C., Valderas, M.S.: Investigación clínica y consentimiento informado en época de pandemia COVID-19: una visión desde la ética de la investigación (Clinical research and informed consent in times of the COVID-19 pandemic: a perspective from research ethics). Med. Clin. **157**(9), e307 (2020). https://doi.org/10.1016/j.medcli.2020.09.005

Bassi, A., Arfin, S., Joshi, R., et al.: Challenges in operationalising clinical trials in India during the COVID-19 pandemic. Lancet Glob. Health **10**(3), e317-319 (2022). https://doi.org/10.1016/S2214-109X(21)00546-5

Bermúdez, J.M.A., Maldonado, L.M.P.: La ética en la publicación científica en tiempos de COVID-19 (Ethics in scientific publishing in times of COVID-19). Revista de Filosofía **38**(99), 225–240 (2021). https://doi.org/10.5281/zenodo.5644537

Bierer, B.E., White, S.A., Barnes, J.M., Gelinas, L.: Ethical challenges in clinical research during the COVID-19 pandemic. J. Bioeth. Inq. **17**(4), 717–722 (2020). https://doi.org/10.1007/s11673-020-10045-4

Bompart, F.: Ethical rationale for better coordination of clinical research on COVID-19. Res. Ethics **16**(3–4), 1 (2020). https://doi.org/10.1177/1747016120931998

Bucci, E., Andreev, K., Björkman, A., et al.: Safety and efficacy of the Russian COVID-19 vaccine: more information needed. Lancet **396**(10256), e53 (2020). https://doi.org/10.1016/S0140-6736(20)31960-7

Bugarín-González, R., Romero-Yuste, S.M., López-Vázquez, P.M., et al.: Experiencia de un comité de ética de la investigación durante la pandemia por COVID-19 (Experience of a research ethics committee during the COVID-19 pandemic). Revista Española de Salud Pública **94**, e202011146 (2020). https://ojs.sanidad.gob.es/index.php/resp/article/view/852. Accessed 2 Mar 2025

Chappell, R.Y., Singer, P.: Pandemic ethics: the case for risky research. Res. Ethics **16**(3–4), 1–8 (2020). https://doi.org/10.1177/1747016120931920

Chaturvedi, S., Kumar, N., Tillu, G., et al.: AYUSH, modern medicine and the Covid-19 pandemic. Indian J. Med. Ethics **5**(3), 191–195 (2020). https://doi.org/10.20529/ijme.2020.058

Cheng, J., Wang, M., Xiao, S., et al.: Ethical issues and countermeasures of clinical research related to outbreak of infectious diseases. Chin. J. New Drugs **29**(24), 2774–2780 (2020). (in Chinese)

Choi, E.K.: Ethical debates surrounding the development of vaccines during COVID-19 pandemic. Bio. Ethics Policy **4**(2), 1–18 (2020). (in Korean). https://doi.org/10.23183/konibp.2020.4.2.001

Christian, A.: Scheitern in der Medikamentenforschung: Zur Bedeutung von Forschungsskandalen für die Entwicklung von Selbstkorrekturmechanismen in der Wissenschaft (Failure in drug research: on the significance of research scandals for the development of self-correcting mechanisms in science). In: Jungert, M., Schuol, S. (eds.) Scheitern in den Wissenschaften: Perspektiven der Wissenschaftsforschung. Brill mentis, Paderborn, pp. 209–236 (2022). https://doi.org/10.30965/9783969752487_011

Cohen, J.: Russia's approval of a COVID-19 vaccine is less than meets the press release. Science, 11 August (2020). https://www.science.org/content/article/russia-s-approval-covid-19-vaccine-less-meets-press-release. Accessed 2 Mar 2025

Cormacain, R.: Parliamentary scrutiny of coronavirus lockdown regulations: a rule of law analysis. Br. Inst. Int. Comp. Law (2020). https://binghamcentre.biicl.org/publications/parliamentary-scrutiny-of-coronavirus-lockdown-regulations-a-rule-of-law-analysis. Accessed 2 Mar 2025

Dadalto, L., Royo, M.M., Costa, B.S.: Bioética e integridad científica en la investigación clínica sobre covid-19 (Bioethics and scientific integrity in clinical research on covid-19). Revista Bioética **28**(3), 418–425 (2020). https://doi.org/10.1590/1983-80422020283402

Ding, W., Mao, S., Ni, S.: Exploring the focus of initial ethical review of novel coronavirus research projects. Chin. J. Clin. Pharmacol. **36**(07), 933–936 (2020). (in Chinese)

Drummond, O.: Human rights challenges during sudden global crisis (2023). https://prepared-project.eu/wp-content/uploads/2023/12/PREPARED-Human-Rights-Report-Final-1.pdf. Accessed 2 Mar 2025

Ehrlich, H., McKenney, M., Elkbuli, A.: Protecting our healthcare workers during the COVID-19 pandemic. Am. J. Emerg. Med. **38**(7), 1527–1528 (2020). https://doi.org/10.1016/j.ajem.2020.04.024

Espinoza-Navarro, O., Rivera-Gutiérrez, S.: Ética y jurisprudencia administrativa de los derechos de los sujetos de investigación en pandemia (COVID-19): función de los comités éticos científicos: Chile (Ethics and administrative jurisprudence of the rights of subjects of research in pandemic (COVID-19): role of the scientific ethics committees: Chile. Int. J. Morphol. **39**(3), 785–788 (2021). https://doi.org/10.4067/S0717-95022021000300785

Fang, Y., Kim, O.J., Jung, J.: The pandemic of COVID-19 and the ethics of human challenge study. J. Korea Bioethics Assoc. **21**(2), 55–74 (2020). (in Korean). https://doi.org/10.37305/JKBA.2020.12.21.2.51

Faust, A., Sierawska, A., Krüger, K., et al.: Challenges and proposed solutions in making clinical research on COVID-19 ethical: a status quo analysis across German research ethics committees. BMC Med. Ethics **22**(96), 1–11 (2020). https://doi.org/10.1101/2020.08.11.20168773

Flores, R.E.P.: Ética de la investigación científica en la búsqueda de la vacuna contra Covid 19 (Ethics of scientific research in the search for a vaccine against Covid 19). In: Enciso, J.A.G., Martínez, A.G. (eds.) Reflexiones éticas en torno a la ciencia y la tecnología. Universidad Autónoma Metropolitana Unidad Azcapotzalco, Mexico City, pp. 22–26 (2020). https://digitaldcsh.azc.uam.mx/index.php/files/93/Reportes-Sociologia/1499/Reflexiones-eticas-en-torno-a-la-ciencia-y-la-tecnologia.pdf. Accessed 2 Mar 2025

Goldman, R.D., Gelinas, L.: COVID-19 and consent for research: navigating during a global pandemic. Clin. Ethics **16**(3), 222–227 (2021). https://doi.org/10.1177/1477750920971801

Gostin, L.O., Lucey, D., Phelan, A.: The Ebola epidemic: a global health emergency. JAMA **312**(11), 1095–1096 (2014). https://doi.org/10.1001/jama.2014.11176

Heidt, A.: Scientists voice concerns over Russian COVID-19 vaccine study. The Scientist, 11 September 2022. https://www.the-scientist.com/scientists-voice-concerns-over-russian-covid-19-vaccine-study-67926. Accessed 2 Mar 2025

Hirt, J., Rasadurai, A., Briel, M., et al.: Clinical trial research on COVID-19 in Germany: a systematic analysis. F1000Research **10**, 913 (2021). https://doi.org/10.12688/f1000research.55541.1

Horton, R.: Offline: COVID-19 is not a pandemic. Lancet **396**(10255), 874 (2020). https://doi.org/10.1016/S0140-6736(20)32000-6

Hsieh, H.F., Shannon, S.E.: Three approaches to qualitative content analysis. Qual. Health Res. **15**(9), 1277–1288 (2005). https://doi.org/10.1177/1049732305276

Hu, R., Dong, H.: Some ethical issues in the state of exceptional catastrophe due to novel coronavirus pneumonia. Med. Controv. **12**(03), 41–45 (2021). (in Chinese)

Inguaggiato, G., Pallise Perello, C., Verdonk, P., et al.: The experience of women researchers during the COVID-19 pandemic: a scoping review. Res. Ethics **20**(4), 780–811 (2024). https://doi.org/10.1177/17470161241231268

Jamrozik, E., Selgelid, M.J.: History of human challenge studies. In: Jamrozik, E., Selgelid, M.J. (eds.) Human Challenge Studies in Endemic Settings . SpringerBriefs in Ethics, pp. 9–23. Springer, Cham (2021). https://doi.org/10.1007/978-3-030-41480-1_2

jmichel2you: Professeur Didier Raoult hydroxychloroquine février 2020 et vaccins février 2022. YouTube (2020). https://youtu.be/cuB0NO5piE8. Accessed 2 Mar 2025

Jung, J., Kim, O.J.: Research ethics in COVID-19 (SARS-CoV-2) vaccine study. Asia Pac. J. Health Law Ethics **13**(3), 1–25 (2020). https://doi.org/10.38046/apjhle.2020.13.3.001

Kadakia, K.T., Beckman, A.L., Ross, J.S., Krumholz, H.M.: Leveraging open science to accelerate research. N. Engl. J. Med. **384**(17), e61 (2021). https://doi.org/10.1056/nejmp2034518

Kadam, A.V., Sandip, P., Suvarna, S., et al.: Challenges faced by ethics committee members in India during COVID-19 pandemic: a mixed-methods exploration. Indian J. Med. Res. **155**(5&6), 461–471 (2022). https://doi.org/10.4103/ijmr.ijmr_1095_22

Kimani, J., Adhiambo Odhiambo, J.: Community consultation with key populations (2023). https://prepared-project.eu/wp-content/uploads/2025/01/PREPARED-Nairobi-workshop.pdf. Accessed 2 Mar 2025

Konstantinoudis, G., Cameletti, M., Gómez-Rubio, V., et al.: Regional excess mortality during the 2020 COVID-19 pandemic in five European countries. Nat. Commun. **13**(482), 1–11 (2022). https://doi.org/10.1038/s41467-022-28157-3

Larousserie, D.: Pourquoi une information judiciaire a été ouverte sur l'IHU de Marseille de Didier Raoult (Why a judicial investigation was opened into Didier Raoult's IHU in Marseille). Le Monde, 6 September 2022. https://www.lemonde.fr/societe/article/2022/09/06/une-information-judiciaire-ouverte-sur-l-ihu-de-marseille-de-didier-raoult_6140376_3224.html. Accessed 2 Mar 2025

Lee, K.: Ethics of human challenge trial for COVID-19 vaccine development: how can it be justified? Asia Pac. J. Health Law Ethics **13**(3), 27–60 (2020)

Leisinger, K., Schroeder, D.: Project lightspeed: a case study in research ethics and accelerated vaccine development. Res. Ethics **20**(4), 847–856 (2024). https://doi.org/10.1177/17470161241251597

Liu, D., Zhang, S., Zhou, J.: Conflict of interest management in clinical studies of novel coronavirus pneumonia. Chin. Med. Ethics **33**(10), 1185–1192 (2020). (in Chinese)

López, I.G.: ¿Pertinencia? del principalismo para la ética de la investigación biomédica regional en tiempos de la pandemia Covid-19 (Relevance of principalism for the ethics of regional biomedical research in times of the Covid-19 pandemic). AlfaPublicaciones **3**(4.1), 26–39 (2021). https://doi.org/10.33262/ap.v3i4.1.122

Mallapaty, S.: Omicron-variant border bans ignore the evidence, say scientists. Nature, 2 December 2021. https://www.nature.com/articles/d41586-021-03608-x. Accessed 2 Mar 2025

Manchola-Castillo, C.: Por una ética en investigación con seres humanos no colonizada desafíos y oportunidades frente a la pandemia por COVID-19 (Towards ethics in research with non-colonised human beings: challenges and opportunities in the face of the COVID-19 pandemic). HOLOS **3** (2022). https://www2.ifrn.edu.br/ojs/index.php/HOLOS/article/view/12878. Accessed 2 Mar 2025

Marzouk, D., Sharawy, I., Nakhla, I., et al.: Challenges during review of COVID-19 research proposals: experience of faculty of medicine, Ain Shams university research ethics committee. Egypt. Front. Med. **2**(8), 715796 (2021). https://doi.org/10.3389/fmed.2021.715796

Mbunge, E.: Effects of COVID-19 in South African health system and society: an explanatory study. Diabetes Metab. Syndr. **14**(6), 1809–1814 (2020). https://doi.org/10.1016/J.DSX.2020.09.016

Mendoza, Y., Abreu, M.: Ética en la investigación clínica en tiempos de covid-19 (Ethics in clinical research in times of covid-19). Paper presented at Jornada 30 Aniversario del Centro Nacional Coordinador de Ensayos Clínicos (CENCEC) (2021). https://cencec30aniv.sld.cu/index.php/cencec/30aniv/paper/view/54/12. Accessed 2 Mar 2025

Miller, J.T.Ö., Türeci, O., Şahin, U.: The Vaccine: Inside the Race to Conquer the COVID-19 Pandemic. St Martin's Press, New York (2022)

Miller, K., Valeva, M., Prieß-Buchheit, J.: Ist wissenschaftliche Integrität geschlechtsneutral (Is scientific integrity gender-neutral)? In: Miller, K., Valeva, M., Prieß-Buchheit, J. (eds.) Verlässliche Wissenschaft: Bedingungen, Analysen, Reflexionen. WBG Academic, Darmstadt, pp. 86–107 (2022)

Mittal, N., Medhi, B.: The bird flu: a new emerging pandemic threat and its pharmacological intervention. Int. J. Health Sci. (Qassim) **1**(2), 277–283 (2007). https://pmc.ncbi.nlm.nih.gov/articles/PMC3068632/. Accessed 2 Mar 2025

Mittermeier, S.: #ichbinhanna: und jetzt (#iamHanna: and now)? De Gruyter Conversations, 30 June 2021. https://blog.degruyter.com/ichbinhanna-und-jetzt/. Accessed 2 Mar 2025

Mlotshwa, L., Andanda, P., Partington, H.: Unprepared: a scoping review of the experiences of health and social care workers in South Africa during COVID-19 (2023). https://prepared-project.eu/wp-content/uploads/2024/11/PREPARED-Experiences-of-the-health-care-and-social-sector-during-COVID-final.pdf. Accessed 2 Mar 2025

Moodley, K., Cengiz, N., Domingo, A., et al.: Ethics and governance challenges related to genomic data sharing in Southern Africa: the case of SARS-CoV-2. Lancet Glob. Health **10**(12), e1855–e1859 (2022). https://doi.org/10.1016/s2214-109x(22)00417-x

Park, Y.S., Kim, O.J.: Government initiatives for research ethics during COVID-19 pandemic in Korea. J. Korean Med. Sci. **39**(12), e116 (2024). https://doi.org/10.3346/jkms.2024.39.e116

Partington, H., Chatfield, K.: Battling at the front line: the experiences of frontline health care staff in the UK during the coronavirus pandemic (2023a). https://prepared-project.eu/wp-content/uploads/2024/11/PREPARED-UK-frontline-personnel-final.pdf. Accessed 2 Mar 2025

Partington, H., Chatfield, K.: Pushed closer to the edge: how the COVID-19 pandemic contributed to the increasing marginalisation of disabled people and their carers in the UK (2023b). https://prepared-project.eu/wp-content/uploads/2024/11/PREPARED-Experiences-of-disabled-people-in-the-pandemic-Final-.pdf

Pearson, H.: How Covid broke the evidence pipeline. Nature, 12 May 2021. https://www.nature.com/articles/d41586-021-01246-x. Accessed 2 Mar 2025

Schroeder, D., et al.: Vulnerability among the Nairobi sex workers, and undertaking community-led research without collecting personal data. In: Schroeder, D., et al. (eds.) Vulnerability Revisited. SpringerBriefs in Research and Innovation Governance, pp. 73–96. Springer, Cham (2024). https://doi.org/10.1007/978-3-031-57896-0_4

Schveitzer, M.C., Thome, B.D.C.: Ética de investigación y asignación de recursos en tiempos de covid-19 (Research ethics and resource allocation in times of covid-19). Revista Bioética **29**(1), 21–26 (2021). https://doi.org/10.1590/1983-80422021291442

Seedall, C., Tambornino, L.: Research ethics and integrity in the DACH region during the COVID-19 pandemic: balancing risks and benefits under pressure. Res. Ethics **20**(4), 650–668 (2024). https://doi.org/10.1177/17470161241229207

Shakespeare, T., Ndagire, F., Seketi, Q.E.: Triple jeopardy: disabled people and the COVID-19 pandemic. Lancet **397**(10282), 1331–1333 (2021). https://doi.org/10.1016/S0140-6736(21)00625-5

Shekhani, S., Saima, I., Aamir, J.: Adapting the ethical review process for COVID-19 research: reviewers' perspectives from Pakistan. East. Mediterr. Health J. **27**(11), 1045–1051 (2021). https://doi.org/10.26719/emhj.21.053

Shin, H.Y.: COVID_19 and IRB review. J. KAIRB **2**(2), 33–37 (2020). (in Korean). https://www.kairb.org/bbs/skin/journal/download.php?code=journal&number=2361. Accessed 2 Mar 2025

Shinomiya, N., Minari, J., Yoshizawa, G., et al.: Reconsidering the need for gain-of-function research on enhanced potential pandemic pathogens in the post-COVID-19 era. Front. Bioeng. Biotechnol. **10**, e966586 (2022). https://doi.org/10.3389/fbioe.2022.966586

Sutrop, M., Parder, M.L., Juurik, M.: Research ethics codes and guidelines. In: Iphofen, R. (eds.) Handbook of Research Ethics and Scientific Integrity, pp. 1–23. Springer, Cham (2020). https://doi.org/10.1007/978-3-319-76040-7_2-1

Swazo, N.K.: Engaging the normative question in the H5N1 avian influenza mutation experiments. Philos. Ethics Humanit. Med. **8**(1), 12 (2013). https://doi.org/10.1186/1747-5341-8-12

Tamariz, L., Hendler, F.J., Wells, J.M., et al.: A call for better, not faster, research ethics committee reviews in the Covid-19 era. Ethics Hum. Res. **43**(5), 42–44 (2021). https://doi.org/10.1002/eahr.500104

Tambornino, L., Lanzerath, D.: Covid-19 human challenge trials: what research ethics committees need to consider. Res. Ethics **16**(3–4), 1–11 (2020). https://doi.org/10.1177/1747016120943635

Taschwer, K.: Wie Covid-19 die Wissenschaft umkrempelte (How Covid-19 changed science). Der Standard, 30 April 2022. https://www.derstandard.de/story/2000135273190/wie-covid-19-die-wissenschaft-schockierte. Accessed 2 Mar 2025

Todhunter, C.: India, COVID and need for scientific integrity. New Age, 8 May 2021. https://www.newagebd.net/article/137404/india-covid-and-need-for-scientific-integrity. Accessed 2 March 2025

Weijer, C.: COVID-19 human challenge trials and randomized controlled trials: lessons for the next pandemic. Res. Ethics **20**(4), 636–649 (2024). https://doi.org/10.1177/17470161231223594

WHO: WHO guidelines on ethical issues in public health surveillance. World Health Organization, Geneva (2017). https://iris.who.int/bitstream/handle/10665/255721/9789241512657-eng.pdf. Accessed 2 Mar 2025

Wilhelmy, S., Müller, R., Gross, D.: Identifying the scope of ethical challenges caused by the Ebola epidemic 2014–2016 in West Africa: a qualitative study. Philos. Ethics Humanit. Med. **17**(1), 16 (2022). https://doi.org/10.1186/s13010-022-00128-y

Wu, Q., Dong, P., Xu, X., et al.: Challenges of ethical review in the development of novel coronavirus vaccines and new drugs. Drug Biotechnol. **28**(4), 400–404 (2021). (in Chinese)

Yoo, S., Kim, E.: Ethical considerations of the researcher and institutional review board on disaster research. Korean J. Ethics **10**(1), 69–98 (2021). https://doi.org/10.38199/KJE.10.4

Zhang, H.: Ethical thinking on the research of public health emergencies based on the prevention and control of pneumonia in Covid-19. China Med. Ethics **33**(04), 415–418 (2020). (in Chinese)

Zhu, W., Yan, F., Zhu, J., et al.: Ethics and integrity challenges during COVID-19 in China. Res. Ethics **20**(4), 683–700 (2024). https://doi.org/10.1177/17470161241245327

Open Access This chapter is licensed under the terms of the Creative Commons Attribution 4.0 International License (http://creativecommons.org/licenses/by/4.0/), which permits use, sharing, adaptation, distribution and reproduction in any medium or format, as long as you give appropriate credit to the original author(s) and the source, provide a link to the Creative Commons license and indicate if changes were made.

The images or other third party material in this chapter are included in the chapter's Creative Commons license, unless indicated otherwise in a credit line to the material. If material is not included in the chapter's Creative Commons license and your intended use is not permitted by statutory regulation or exceeds the permitted use, you will need to obtain permission directly from the copyright holder.

From Real-World Challenges to a Global Code: How the PREPARED Code Was Built

Natalie Evans[1](✉), Hazel Partington[2], Doris Schroeder[2,3], Kate Chatfield[2], Clàudia Pallisé Perelló[1], Nandini Kumar[4], Ock-Joo Kim[5], Wei Zhu[6], and Dafna Feinholz[7]

[1] Department of Ethics, Law and Humanities, Amsterdam UMC, Vrije Universiteit Amsterdam, Amsterdam, The Netherlands
n.evans@amsterdamumc.nl
[2] Centre for Professional Ethics, University of Central Lancashire, Preston, UK
[3] School of Law, UCLan Cyprus, Larnaca, Cyprus
[4] Forum for Ethics Review Committees in India, New Delhi, India
[5] Seoul National University, Seoul, South Korea
[6] Fudan University, Shanghai, China
[7] United Nations Educational, Scientific and Cultural Organization, Paris, France

Abstract. The PREPARED Code is a risk-based, values-driven framework that integrates research ethics and research integrity and is designed for a global audience. Developed over two years, this ambitious initiative required a collaborative, multidisciplinary effort led by an international team. The PREPARED team employed a range of methods to develop the code, including literature searches, scoping reviews, empirical studies, targeted consultations, ethical and legal analyses, and public consultation. This chapter explores the processes and methods used to develop the PREPARED Code, highlighting how real-world challenges in research ethics and research integrity during crises were identified, analysed and validated by stakeholders. It describes how these challenges were aligned with universally recognised moral values and grouped as risks, and how the risks were transformed into a clear, focused and jargon-free code of conduct. It also details the final stages of development, which involved iterative refinement of the code from Version 1 to Version 13, through extensive consultation and review.

Keywords: Research ethics · research integrity · risk-based · code of conduct

1 Introduction

The PREPARED Code was envisioned as an operational ethics and integrity framework to facilitate a swift and effective research response during pandemics while upholding key ethical values. Planned to be applicable across all research disciplines, combining both research ethics and research integrity, values-driven, and suitable for a global audience, it was an ambitious endeavour requiring a collaborative and multifaceted approach.

In September 2022, a dedicated team of 16 partner institutions and 14 specialist advisers from five continents set out to develop the PREPARED Code: A Global Code of Conduct for Research during Pandemics. Over the next two years, the team conducted literature searches, scoping reviews, empirical studies, targeted consultations, ethical and legal analyses, and a public consultation, culminating in the completion of the code on 31 December 2024.

The work was guided throughout by a carefully designed rationale (see Chap. 3), much of which had been tested during the development of the TRUST Code (TRUST 2018), an ethics code for equitable research partnerships. In fact, the guiding rationale for the development of the PREPARED Code closely mirrored that used for the TRUST Code (Schroeder et al. 2019), including the bottom-up, risk-based, values-driven and inclusive approach (see Chap. 3). However, while guided by a similar rationale, the methods that were implemented for the development of the PREPARED Code differed from those of the TRUST Code as they needed to be tailored to the pandemic context.

This chapter describes the steps taken and the methods employed by the PREPARED team for the development of the PREPARED Code. We first explain how evidence of real-world research ethics and research integrity challenges was gathered, analysed and validated by stakeholders. We then clarify how the challenges were themed and mapped onto globally understandable moral values. The remainder of the chapter details how the PREPARED Team moved from Version 1 of the PREPARED Code to Version 13 through extensive and inclusive consultations.

2 The PREPARED Methods: An Overview

The process of developing the PREPARED Code was shaped by a clear rationale or methodological approach as described in Chap. 3. The methodological approach (risk-based, values-driven, etc.) determined the overall strategy for development, but there were many ways in which the strategy could have been implemented. In other words, there were many different methods or "procedures, tools and techniques" (Schwandt 2001: 158) that could have been used to collect and analyse data to inform the development of the PREPARED Code.

For high-quality research, the selection of appropriate methods and procedures must be tailored to the context in which the activities are taking place (Jansen et al. 2010). Additionally, the methods must be consistent with the overall methodological approach (Wright et al. 2016). Table 1 provides an overview of how the project activities were tailored to reveal the research ethics and research integrity challenges relevant to the pandemic context, while remaining aligned to the guiding rationale for the development of the code.

The implementation activities listed in Table 1 were undertaken in a series of steps that flowed from the identification of research ethics and research integrity challenges during pandemics through to the refinement of the PREPARED Code as shown in Fig. 1.

In the following sections, each of these steps is described further to show how they were undertaken.

Table 1. Alignment of the guiding rationale (methodology) with activities undertaken during development of the PREPARED Code

Guiding methodological factor	Implementation activities
The PREPARED Code is built on real-world risks	• Literature reviews on research ethics and research integrity challenges during COVID-19 in nine languages • Scoping reviews on research ethics and research integrity challenges during avian flu and Ebola epidemics in English • Literature-based human rights analysis • Empirical and literature-based studies to reveal general challenges for groups in vulnerable situations • Validation workshops to check the identified challenges
The PREPARED Code is values-driven	• Values mapping of the challenges to the four values framework of the TRUST Code • Investigation to identify value gaps, e.g. solidarity?
Research ethics and research integrity are integrated in a unified code	• Literature reviews on research ethics and research integrity challenges during COVID-19 in nine languages • Scoping reviews on research ethics and research integrity challenges during avian flu and Ebola epidemics in English • Validation workshops to check the identified challenges
A broad and inclusive approach to development was taken	• Empirical and literature-based studies to reveal general challenges for groups in vulnerable situations • Creation and involvement of stakeholder platforms for broad code consultation and validation events • Analysis of pandemic/crisis guidance documents to ascertain whether the risk analysis had possibly overlooked any major challenges

3 Gathering Evidence of Real-World Challenges

Fundamental to both the PREPARED Code and the TRUST Code (TRUST 2018) is that they address all major real-world risks. For the PREPARED Code this meant a focus on pandemics and for the TRUST Code a focus on equitable international research collaborations. First and foremost, those risks had to be identified.

For the TRUST Code, this entailed extensive consultation and searching for real cases of inequitable research collaborations, because such cases were not well represented in

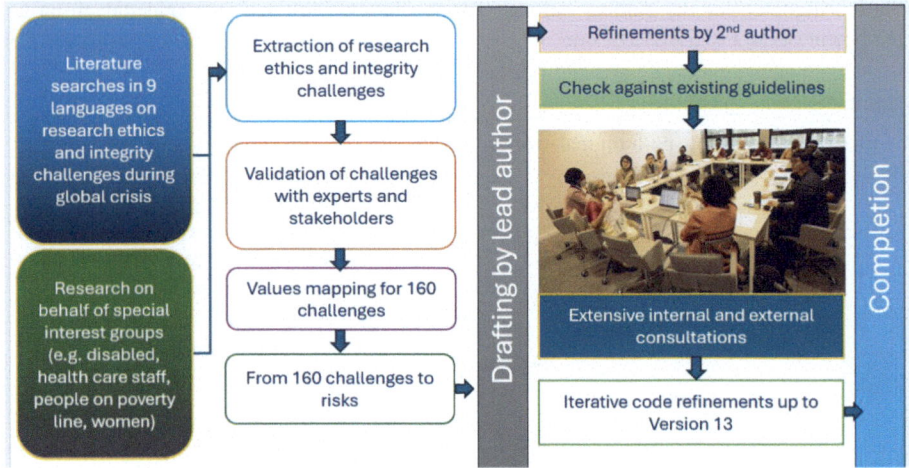

Fig. 1. Steps in the development process of the PREPARED Code

the literature. It was even necessary to launch a case study competition to uncover examples of unethical research partnerships (Schroeder et al. 2018).

The situation was very different for the PREPARED Code. COVID-19 saw an explosion in global publishing related to the pandemic (Fassin 2021). The PREPARED team was able to tap directly into this rich body of evidence to identify the research ethics and research integrity challenges encountered during the pandemic. How evidence was gathered is fully described in Chap. 4 and summarised in Table 1 as follows:

- literature reviews on research ethics and integrity challenges during COVID-19 in nine languages
- scoping reviews on research ethics and integrity challenges during avian flu and Ebola epidemics in English
- literature-based human rights analysis
- empirical and literature-based studies to reveal general challenges for groups in vulnerable situations.

The extensive research work spelled out above, undertaken simultaneously in Europe, Africa and Asia, produced a vast amount of rich data for analysis, which began with the extraction and initial sorting of the research ethics and research integrity challenges.

4 Extraction of Research Ethics and Research Integrity Challenges

The core analysis team was made up of three people: Hazel Partington and Kate Chatfield (referred to here as the "analysts") and the lead author of the PREPARED Code, Doris Schroeder (referred to here as the "lead author"), who also acted as quality controller throughout.

To ensure that the challenges for research ethics and research integrity were extracted consistently, this was initially done by one person (the lead author), who tabulated the identified challenges in an Excel spreadsheet with one sheet per language.

Following tabulation, the two analysts sorted the challenges into those related directly to research ethics, to research integrity, and to broader "context-specific" challenges that were not specifically related to research (e.g. shortages of personal protective equipment for healthcare staff during a pandemic). Research ethics and research integrity challenges were then inventoried on new Excel spreadsheets that listed each specific real-world research challenge, the reference or source describing it, and which language report it had been identified in.

The resulting tables per language of the original research were checked by the research authors. For example, South Korean colleagues checked that the identified challenges matched those in the Korean language report. Table 2 provides some examples by way of illustration.

Table 2. Example of ethics and integrity challenges for research during pandemics

What happened in the real world?	Reference/source	Which report?
Interviews and focus groups were switched to digital	Bartmann et al. (2022)	German
Outrage erupted at alleged "ethics dumping" after French doctors said that COVID-19 studies should be carried out in Africa, where there was less virus protection	Le Monde with AFP (2020)	French
The Sputnik vaccine's efficacy and safety were allegedly announced before clinical trial completion	Cohen (2020)	Russian
In a multicentric trial involving 42 sites, the decision of the Central Ethics Committee was followed at only three sites	Bassi et al (2022)	Hindi*
Uncoordinated, low-powered studies were conducted in multiple locations	Jung and Kim (2020)	Korean*
Healthcare providers had insufficient time to collect the follow-up data on patients necessary for study completion	Liu et al. (2020)	Chinese
The use of online platforms with weak security features raised concerns about potential breaches of confidentiality	Ghooi (2020)	English
Pre-prints and fast-tracked publications decreased scientific rigour and increased the number of publication retractions	Bermúdez and Maldonado (2021); Dadalto et al. (2020)	Spanish

* *The Hindi and Korean literature review also included items reported in English about India and South Korea, given that English is an important language of scholarly communication regarding national research ethics matters in those countries.*

Following the sorting and checking, and the removal of context-specific challenges, a total of 160 research ethics and research integrity challenges were identified.

Together, the findings from these reviews provided a detailed and inclusive mapping of global research ethics and research integrity challenges. Since all the challenges were extracted from real-life cases, they offered a representative and nuanced foundation for the development of an ethics code that could be globally relevant while taking special account of groups in vulnerable situations.

5 Validation of Challenges by Stakeholders

In the next step of the development process, the experiences and perspectives of stakeholders (including experts) who had faced research ethics and research integrity challenges in practice were explored to confirm that their insights were in line with what had been found in the literature and empirical studies. The PREPARED team also hoped to uncover any additional challenges that had not yet been identified.

To this end, online focus group discussions, called "validation workshops", were convened. These workshops brought together the stakeholders who had experienced research ethics and integrity challenges first-hand, or who could speak with authority on behalf of the groups or networks they represented. Four separate online workshops were conducted with research policymakers, ethics and integrity experts, senior researchers from various disciplines, and representatives from disease-specific European advocacy groups.

Experts were recruited from established networks, including the European & Developing Countries Clinical Trials Partnership (EDCTP), the European Network of Research Ethics Committees (EUREC), the European Network of Research Integrity Offices (ENRIO) and pan-European advocacy groups, aiming for a balance of expertise and diverse perspectives.

Each workshop began with an introduction to the PREPARED project, followed by a presentation of the key research ethics and research integrity challenges that had already been identified. During the discussions, facilitated by Natalie Evans, the stakeholders highlighted challenges specific to their group and how they thought these challenges might be addressed via the PREPARED Code. Some illustrative examples of input per group are given below.

Policy and research ethics experts emphasised the need for practical operational guidance for research ethics committees during health crises. They discussed the importance of good communication across decision-making levels and clearer guidance on issues like online consent and multisite trial adjustments. They stressed challenges in returning to normal procedures post-pandemic and the need for additional resources and innovative training. They also highlighted justice considerations, particularly fair benefit-sharing for low- and middle-income countries.

> For the sake of fairness and justice, I think it would be good to demand that researchers address human rights and human dignity because the question usually would be: what is the significance of any study that is being conducted and what are the possible risks and benefits? Researchers trying to deal with this would be looking at it from the lenses of human rights and human dignity so that ethical considerations would be made for individual participants and the public.
>
> Dr Lillian Omutoko, Associate Professor, University of Nairobi and National Bioethics Committee Member

Research integrity experts highlighted the fact that the pandemic had exacerbated existing research integrity challenges, but also accelerated the adoption of solutions like open data and living reviews (systematic reviews that are continually updated with new relevant evidence). Transparency issues, data-sharing barriers, a lack of coordination and collaboration between sectors, and difficulties communicating science to the public were discussed. Improved public communication and transparency about uncertainties were seen as critical for building trust.

> There was such a huge gap between how researchers talk about what they're doing, how it's communicated within science, and how the public at large understands this communication. Or rather doesn't understand it at all and feels that this is all very uncertain and can't be trusted…. Trust in a very important institution, science, was eroded. There's no easy solution to that.
>
> Sabine Chai, Managing Director, Austrian Agency for Research Integrity

Researchers from diverse disciplines discussed which knowledge had been prioritised in the pandemic policy response, describing the neglect of attention in pandemic policymaking to some disciplines, such as the social sciences and economics. Researchers also described the negative effects of rushing proposals to chase pandemic funding, and of lockdown measures on the quality of data collection and the training of the next generation of researchers. Like the research integrity experts, they also described the pandemic as exacerbating existing problems within academia and emphasised the need to strengthen research support structures in preparation for the next crisis.

Expert representatives of European advocacy groups reported the difficulties their members had in understanding the language used to communicate scientific information. Patients and individuals living with pre-existing conditions often felt alone in evaluating their specific risks in relation to treatments and vaccines. In clinical settings, there was also a blurring of the line between treatment and research, and a lack of options apart from participation in research.

Experts from all workshops also offered advice for the drafting of the PREPARED Code. This included a recommendation that the code should not have a preamble describing the challenges that had been faced more broadly during the pandemic but could not be addressed by the main target audiences of the PREPARED Code, that is, researchers, research ethics committees or research integrity offices (see Sect. 8.5).

> To me, you need to make clear what the code is not about as well. A code which is about everything is useless. It means nothing anymore. You need to be really clear about what you're not talking about, and what you're not giving guidance on, and that might be a good content of the preamble. Not a preamble saying, "Hey, we needed to do that and that and that and we couldn't bring it in, so here it is." That doesn't make sense to me.
>
> Lex Bouter, Professor Emeritus of Methodology and Integrity, Vrije Universiteit Amsterdam.

After the workshops, summaries of the main themes were compiled from each session and sent to the participants in the form of a "validation workshop report" to ensure that discussions were captured accurately.

6 Values Mapping

Just like the TRUST Code, the PREPARED Code was intended to be values-based, so that the specific recommendations for research ethics and research integrity were linked to commonly understood moral values. Such a values-based approach creates a strong connection between *what* should be done during pandemics and *why* (morally) it should be done (Schroeder et al. 2019). Nevertheless, while the TRUST values of fairness, respect, care and honesty had resonated globally, their applicability to the PREPARED Code could not be taken for granted. Until the wide-scale research ethics and research integrity risks encountered during pandemics were identified, alignment of the challenges with the TRUST values was purely a matter of speculation.

As a starting point, the two analysts used the four TRUST values as a deductive framework for the analysis. They coded the research ethics and research integrity challenges independently: for each challenge they decided which of the four moral values was most at risk of being violated. The challenges to research ethics and research integrity that might be associated with more than one moral value were organised under the primary moral value at stake. To give the reader an idea of what this process looked like, here is an example.

A challenge from the PREPARED English-language report on research ethics and integrity challenges during COVID-19 was described as follows: "Researchers had to rely on ICU nurses and doctors to follow up enrolled participants on their behalf and share monitoring reports since they were not allowed to enter the ICUs."

To identify the values that this challenge illuminated, the analysts had to decide which moral values were being compromised or violated when researchers and ICU staff found themselves in these situations. In this case, it could be argued that both care and fairness were implicated. It was necessary for ICU nurses and doctors to collect data directly in order to protect patients and researchers from infection. Yet the additional workload and pressure on ICU staff could lead to stress and exhaustion, constituting a violation of the value of *care*. The same additional burden could also be interpreted as a violation of the value of *fairness*.

During this stage of the analysis, it was vital to ensure that the analysis remained grounded in the data to assess which was the main value at stake. For this example, both analysts deemed "fairness" to be the most important value at stake, due to the unfair burdens of data collection on ICU healthcare staff. Any disagreements between analysts were resolved through discussion with input from the lead author.

Additionally, the analysts remained open to the possibility that some challenges might be related to different moral values. For example, the moral value of *solidarity* has been described as important in guiding a global pandemic response (Dawson et al. 2020; Tomson et al. 2021), and it was reasonable to expect that solidarity might be required in a moral values framework that governs pandemics. However, while the relevance of solidarity to a small number of the risks was evident, these risks were deemed primarily matters of fairness and/or care. In fact, some scholars and commentators view solidarity and fairness as two closely related moral values of the same group, rather than clearly distinct entities (Küçük 2016; European Commission 2020; Cappelen et al. 2021).

Further, given the inclusion of research integrity challenges, more specific research integrity-related values such as accountability (ALLEA 2017) were also considered. But accountability did not represent the *main* value at stake for any risk identified from the real-world challenges; there were no specific risks related to accountability that were not already represented by the values of honesty and fairness. While accountability might not be intuitively understood as falling under these values, it is contingent upon the honesty of the person being held to account and *may also involve some type of justice or fairness*[1] (Chatfield and Law 2024).

Thus, it soon became clear, during the process of mapping values for the PREPARED Code, that *all* of the identified pandemic-related challenges for research ethics and research integrity could be aligned with at least one of the four TRUST values. In other words, the identified breaches of research ethics and research integrity that emerged or were exacerbated during pandemics could all be associated with lapses or failures in fairness, respect, care and/or honesty.

In total, 160 challenges were identified and mapped to the TRUST values. Of these, 39 (24%) related to fairness, 29 (18%) to respect, 74 (46%) to care, and 18 (11%) to honesty (see Fig. 2).

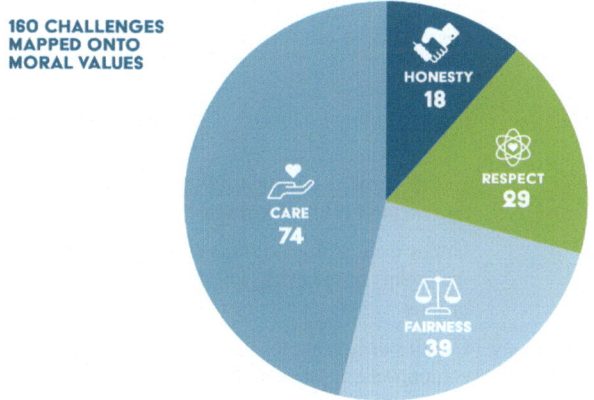

Fig. 2. Challenges mapped onto moral values

7 From Challenges to Risks

The next step of analysis entailed translating the reported challenges into descriptions of the potential *risks* for research ethics and research integrity during pandemics. This step also involved identifying the parties that might be affected by those risks (e.g. research participants, medical staff, research ethics committees and researchers). Again, this was undertaken independently by the two analysts and then compared and agreed through in-depth discussion with the lead author.

---

[1] For instance, distributive justice (fairness in distribution), procedural justice (being treated fairly), retributive justice (a correction or punishment) or restorative justice (to right a wrongdoing).

First, similar challenges were grouped together, for example this challenge from the literature review in English: *Minimising risks was difficult to guarantee due to lack of preliminary data on the investigational agents or approved drugs* (Kadam et al. 2022), and this one from the review in Mandarin: *A project wanted to study an antiviral drug, but action targets for experimental drugs did not exist in coronavirus* (Zhang et al. 2020). Both describe challenges associated with the testing of new interventions for a novel disease. For research ethics, this poses a risk to the consent process because participants should be informed about the potential harms and benefits involved before they decide whether to participate in a study. In other words, during pandemics, there can be a risk to the consent process if there is uncertainty about the disease and/or potential treatments (Article 11 in the PREPARED Code).

Secondly, once the challenges had been grouped, the risks were described in terms relevant to research ethics and/or research integrity, together with the parties that might be affected. For illustrative purposes, Tables 3, 4, 5 and 6 present selected examples of the risks we identified, those affected by the risks and the main moral value at stake in each case.

Table 3. Illustrative examples of *fairness* risks during global health crises

FAIRNESS	
Risks for:	
Research participants	Unfair burdens when participating in poor quality studies that had no possibility of benefit
Society	Unfair exclusion of certain groups from studies meant that there were gaps in the evidence/interventions not tested for these groups
Research ethics committees	Unfair burden due to: • Increased number of studies • Pressure for rapid review/turnaround • Other work pressures (many in healthcare) • Fewer people available to undertake reviews • Switch to alternative ways of working (e.g. online methods) which can be problematic for some
Healthcare staff	Unfair additional burdens for ICU staff who had to help with data collection and monitoring

With the risks to research ethics and research integrity having been identified, it was now possible to start drafting the PREPARED Code.

8 Creating the First Draft

After 18 months of evidence-gathering and analysis, it was time to develop the first draft of the PREPARED Code.

Table 4. Illustrative examples of *respect* risks during global health crises

RESPECT	
Risks for:	
Research participants	Consent issues (a selection): • Research ethics committees did not have the necessary information for evaluation of risks and consent procedures • Consent processes had to be adapted (e.g. proxy and e-consent) with unknown impacts • Consent possibly compromised due to accessibility challenges with very sick patients in isolation
Society	• Many institutions did not comply with reporting and data-sharing obligations • Lack of respect for opinion of experts • Lack of compliance with research ethics norms and requirements
Research ethics committees	Lack of respect for REC authority, opinions and decisions

The first draft was written by the lead author. The initial individual effort allowed for a consistency of voice as had proven beneficial during the development of the TRUST Code (Schroeder et al. 2019). Reducing a large number of specific risks to a smaller number of succinct articles was achieved by applying four steps of synthesis (see Fig. 3):

- focusing on the pandemic context
- tailoring results to target audiences
- grouping the risks so that several could be addressed through one article
- examining the depth of specificity.

8.1 Focusing on the Pandemic Context

Thousands of ethics codes already exist. In fact, the PREPARED team analysed 236 new ethics guidance documents for COVID-19 alone (See Chap. 4). With this proliferation of ethics documents in mind, the PREPARED Code authors aimed to develop a short, jargon-free code tailored to a particular situation, namely the next pandemic. One way of keeping the new code short and focused was to avoid the inclusion of recommendations that were already addressed in other widely adopted ethics and integrity guidance instruments. The PREPARED Code is designed to be complementary to other well-established codes. Indeed, some, like the TRUST Code, are cross-referenced because they are also relevant to pandemic times.

The risk of ethics dumping (the export of unethical research practices from higher- to lower-income countries (Schroeder et al. 2018) was identified in several of the literature reviews for the PREPARED Code. However, recommendations related to ethics dumping are already described in the TRUST Code: A Global Code of Conduct for Equitable Research Partnerships (TRUST 2018). Furthermore, the TRUST Code was developed by a group that consisted, in the main, of teams from low- and middle-income countries,

Table 5. Illustrative examples of *care* risks during global health crises

CARE	
Risks for:	
Research participants	Potential harm from: • Pressured research ethics committees which may not have time for due diligence • Face-to-face interactions (infection risk) • Receiving placebo (in placebo-controlled studies) • Participating in human challenge studies • Data breaches due to modified informed consent collection procedures (e.g. remote digital consent) Unnecessary burdens from: • Lack of coordinated studies • Flawed study designs Potential for therapeutic misunderstanding when rushed during consent process
Society	Reduced trust in science from: • Misinformation and/or sensationalist reporting • Failure to ensure quality and retract questionable publications
Research ethics committees	Potential for harm or stress from: • Pressures to review quickly • Resource shortages • Switch to remote working
Health care personnel	Increased burdens because only they could access participants in ICUs
Animals	Potential for harm if regulatory reviews not carried out or not carried out effectively

Table 6. Illustrative examples of *honesty* risks during global health crises

HONESTY	
Risks for:	
Research participants	• Lower data protection standards in crisis situations • Research participants not informed about use of their data • Patients not informed about collection and use of samples
Society	Promotion of drug based on flawed or unverified information

thus achieving appropriate representation on the topic (Schroeder et al. 2019). It was therefore decided to cross-reference the TRUST Code rather than add guidance articles tackling ethics dumping to the PREPARED Code.

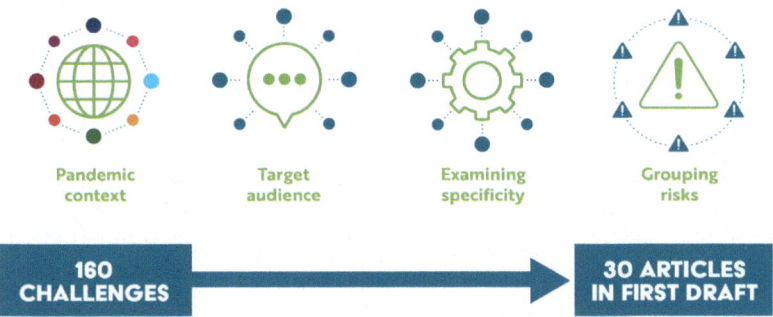

Fig. 3. Four steps of synthesis

8.2 Tailoring to the Target Audience

An ethics code is best aimed at individuals from a defined group; it will help those individuals undertake particular activities ethically through written guidance (Giorgini et al. 2015). The PREPARED Code is primarily aimed at researchers, and secondarily at research ethics committees and research integrity offices. The latter two *assist* researchers in undertaking their research ethically. Hence, they can also benefit from ethics codes in their advisory roles. This meant that some of the risks did not need to be addressed by the code's drafting team, as they were not relevant to these groups.

For example, resource shortages experienced by research ethics committees are an institutional issue that cannot be resolved by researchers alone. The fact that poor-quality publications remain in the published domain due to inaction by (predatory) publishers is not within the realm of researchers' influence (Barrière et al. 2023). Vaccine availability for lower-income settings is also not something researchers can readily address (Schveitzer and Thome 2021). It requires action at international level.

Because ethics deals with messy social worlds, it is not always possible to develop distinct categories, so there are four cases where the PREPARED Code refers to challenges that are not fully within researchers' power. These references were included because researchers can carry some of the responsibility for these aspects, and the PREPARED team decided to promote awareness of them.

First, the lead author added a vision statement to the code to make clear that all code authors believed firmly that questions of global access to vaccines were crucial in pandemic ethics, even though this was not the responsibility of researchers: *"Pandemic research should be trustworthy and the results accessible to all."* In this way the code at least acknowledges prominently the intractable problem of making vaccines accessible to all.

Second, two articles address research ethics committees *directly*. Article 7: *"RECs should expedite the evaluation of research proposals that address urgent societal needs without compromising rigorous ethical standards."* Article 21: *"During pandemics, researchers may experience a heightened risk of hostility and related safety and security concerns. Research ethics committees should check that risk management plans are in place.*" This explicit guidance to research ethics committees was intended to provide an additional level of protection for researchers, in response to ample evidence of heightened safety and security risks to them during COVID-19.

Third, Article 15 stipulates: *"Especially during pandemics, researchers who handle potentially infectious biological materials should be adequately trained and equipped to safeguard public health."* One could argue that getting staff trained is exclusively an institutional responsibility. However, successful training also requires good time management and motivation on the part of employees, and hence it was added to the PREPARED Code as a partial responsibility for researchers.

Fourth, two further articles might be considered beyond the power of researchers to implement: namely, Article 2, on coordinating research and avoiding wastage, and Article 4, on continuing community engagement during a major crisis. Indeed, both require multiparty involvement. Nevertheless, they were included in the code because researchers are not completely powerless in these areas. For instance, collaborating with as many colleagues as possible rather than trying to recruit to multiple small studies is something that researchers can consider. More obviously, successful community engagement is best driven by research teams. Hence, these two articles were included to raise researchers' awareness of the role they can play.

8.3 Grouping Risks

A synthesis step, which reduced the number of potential articles considerably, was the grouping of risks. The effect of this was to consolidate the ten risks relating to informed consent which had been identified in the literature reviews and subsequent values analysis into just three articles focused on consent in the PREPARED Code (Articles 9–11). For the purpose of precision and focus, this smaller number of articles addressed all the risks identified.

8.4 Examining the Depth of Specificity

Several of the literature reviews identified very specific research ethics challenges during a major crisis, for instance the most detailed and regularly cited ethical issues in human challenge studies (see Weijer 2024). The first version of the PREPARED Code included the following article: "During pandemics, healthy volunteers who take part in Stage 1 vaccine trials, carry risks and burdens for humankind. Researchers involved in such studies should follow the separate *Ethics Check List* for First-in-Human Vaccines."

The Ethics Checklist cited was also drafted by the lead author based on substantial work undertaken by the VolREthics Initiative (Inserm 2022). The checklist included 11 precise checkpoints, such as: "In bioconfinement, access to facilities, which counteract feelings of isolation, must be provided to ensure continuous wellbeing (for example, wifi, phones, TV, space, windows)," or "When offered, completion bonuses should be modest." This level of specificity would have been inappropriate for the PREPARED Code.

8.5 Input from the Validation Workshops

Following the consolidation work described above, the lead author formulated short articles in the format of ethics code guidance and checked whether the resulting draft code was compatible with the challenges, risks and suggestions identified in the report from the validation workshops (see Sect. 5).

One important decision had already come out of the validation workshop with research integrity experts. The PREPARED Code has no preamble, but merely a small number of introductory sentences (see Sect. 5). At first the lead author was keen on a preamble to distinguish the broader ethics issues identified in addition to the more specific research ethics and integrity issues. However, it was argued in this validation workshop that a preamble would reduce clarity by conflating different challenges. This idea was therefore dropped in favour of a single-sentence vision statement.

The validation workshops also unearthed one topic that was not raised in any of the nine language reports: benefit-sharing. While the compatibility of COVID-19 virus sharing (samples and genome) with the requirements of the UN Convention on Biological Diversity (CBD) and national biodiversity laws was discussed in the literature (Humphries et al. 2021; Sett et al. 2022), the pandemic *research ethics* literature we reviewed did not mention the topic. And as the CBD only covers non-human genetic resources, the CBD-related literature was not relevant to the coronavirus responsible for COVID-19.

In line with the vision statement, justice considerations formed a major part of ethics discussions during and after the COVID-19 pandemic, as was also emphasised by several delegates at the validation workshops. The lead author therefore agreed that a new article should be included in the code, which is now Article 3: "*A fair plan for access to the benefits of pandemic research should be agreed early on in any project, in collaboration with stakeholders.*"

8.6 Completing the First Draft

Progressing from the identified risks to a draft code of 30 articles took the lead author six weeks. This draft was then checked by the second author, Kate Chatfield, who suggested refinements across all topics. Following that check, it was sent to Natalie Evans for a focus on research integrity, to Pamela Andanda for a focus on Global South applicability and to Joshua Kimani for a focus on the adequate representation of the interests of persons in highly vulnerable situations.

At the same time, the draft including the refinements by the second author (Version 2) was sent to three external advisers, Prof. Fatima Alvarez-Castillo in Manila, Prof. Jantina de Vries in Cape Town and Prof. Charles Weijer in London, Canada. They were kind enough to provide video feedback in advance of the Amsterdam meeting (see Sect. 9.1). Here are examples of changes made in response to useful adviser input:

- The order of articles within each moral value was revisited to align with the steps in the research process.
- The phrase "with adequate protections" was added to Article 5, which deals with the inclusion of persons in vulnerable situations in research.

- The excellent phrasing about communicating risks and benefits "in terms of what is known, what is uncertain and what is unknown" in Article 11 was suggested by Charles Weijer.

9 Broad Consultation and Refinement

The PREPARED Code went through 13 iterations before it was finalised. Going from Version 1 to Version 13 involved disseminating the draft code as widely as possible to gather a wide range of perspectives and feedback.

Consultation formats differed, but all allowed for general feedback and comments on five specific questions:

1. Are the articles clear and understandable?
2. Can the ordering of articles be improved?
3. Is each article under the right value?
4. Do all disciplines feel covered?
5. Have we omitted anything important?

9.1 The Amsterdam Meeting

In its second draft, and accompanied by three videos from external advisers, the code travelled to Amsterdam for the opening of the consultations. At an in-person meeting of the PREPARED team partners and advisers (51 experts) in May 2024, input was collected via group discussions focusing on research ethics, research integrity and global relevance (Fig. 4).

Fig. 4. Group discussing the global relevance of the draft PREPARED Code, Amsterdam 2024

This round of expert feedback helped the lead author refine the code's articles and their relevance and applicability to the target groups of researchers, research ethics committees and research integrity offices. Also addressed were matters of content and format.

For instance, regarding content, there were discussions about what to do with risks that would require action *before* the next pandemic. The team considered the options of providing additional "resilience" or "preparedness" recommendations as part of the code, or of developing and referring to additional preparedness guidance. In the end, the team chose the latter option, to help keep the code short and jargon-free and to make it easier to update additional preparedness resources.

Regarding format, the meeting discussed the order of the values of fairness, respect, care and honesty in the code. Most of the Europeans in the group wanted *care* to be addressed first in the PREPARED Code, while the majority of the global team wanted *fairness* first. The final PREPARED Code starts with the value of fairness.

9.2 Dissemination to External Stakeholders

Wider dissemination of the PREPARED Code to stakeholders started around a month after the Amsterdam meeting, so there was time to refine the code in the light of the suggestions made at the in-person meeting.

The first external groups to be contacted for consultation were those already established via the PREPARED "stakeholder platforms". These had been formed during the time of evidence-gathering, led by consortium partners, to represent important networks of research stakeholders (Fig. 5). The platforms lend PREPARED the credibility, and the global reach, to solicit valuable comments from the right people on continuously refined drafts of the PREPARED Code.

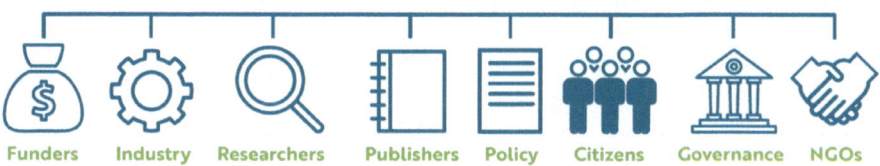

Fig. 5. PREPARED stakeholder platforms

In addition to circulating the code to platform members, the PREPARED team organised a wide range of activities from June to November 2024 to seek feedback from the following groups:

- experts working at the level of research ethics and research integrity policy and practice nationally (e.g. members of EUREC, the Forum for Ethics Review Committees in India and ENRIO) and internationally (e.g. members of the International Bioethics Committee of UNESCO and the World Commission on the Ethics of Scientific Knowledge and Technology)
- experts in specific areas of research ethics (e.g. experts in early-stage clinical trials and senior pharmaceutical industry bioethicists)
- researchers and students in relevant disciplines (e.g. metascientists, law and education researchers, and emergency ethics experts).

The public were also invited via a social media campaign to submit comments on the PREPARED Code via the PREPARED website in October and November 2024.

9.3 Considering and Responding to Comments

Between May and December 2024, the PREPARED Code went through 13 revisions. Most changes were triggered by feedback from the consultation activities, but some arose from further internal work by the PREPARED consortium.

For all potential revisions, the following criteria were applied:

- Proposed changes had to meet the four criteria from the synthesis process outlined above, that is, focusing on the pandemic context, tailoring results to target audiences, avoiding a proliferation of articles by combining issues, and avoiding overly deep specificity.
- Suggestions for changes had to be accompanied by evidence that a real-life challenge was involved.
- Suggestions that might be difficult to apply globally were to be avoided to ensure that the code would be useful around the world.

For consistency in decision-making, the lead author was responsible for the final version of all articles (in collaboration with a professional language editor). However, she convened small, fast-action, often ad hoc groups for many discussions to obtain further input and help her arrive at well-reasoned decisions. These small groups were usually needed after suggestions from external consultations. In addition, all changes were approved by the second author and, in the final instance, by all 57 authors.

10 Examples of Refinements Following Feedback

Below are four concrete examples of refinements arising from different types of consultations.

10.1 Written Consultations Through the Eight PREPARED Stakeholder Engagement Platforms

Several of the consultations with the platforms were undertaken in writing. Some feedback resulted in changes including the following:

- Consultation with industry (bioethics colleagues from Roche and Novartis)

  - The term "promptly" was added to Article 23, which requires researchers to inform participants and research ethics committees of changes in the risks or burdens of participation in clinical research.
  - The term "study suspensions" was replaced with the term "study modifications" in Article 18. This way, the impact on all those who depend on research studies for access to medication and services must be considered during a pandemic, not just the impact on those whose studies have been suspended.
  - The term "deception" was added to Article 27 about public communication by researchers.

- Consultation with the research integrity platform

- The phrase "or their proxies" was added to Article 9 about informed consent, given that not all research participants will be in a situation to make decisions for themselves during a pandemic.

10.2 In-Person Consultations at Conferences or Through Webinars

Some consultations were run as conference presentations or through webinars. At the ENRIO research integrity conference in Prague, an entire session was dedicated to feedback on the PREPARED Code, resulting in changes that included the following:

- The term "study limitations" was added to Article 27, which addresses how and what researchers should communicate publicly. In addition, to reduce jargon, the term "veracity" was removed from the article.
- The phrase "To promote public trust" was removed from Article 26, which asks researchers to answer publishers' research ethics questions. It was regarded as an unfounded deduction.

10.3 Gap Analysis

The risk-based approach of the PREPARED Code (see Chap. 3) demanded that research and consultation input inform every single article of the code. However, the PREPARED team also consulted existing ethics guidance, identifying research ethics and integrity challenges covered in existing COVID-19, Ebola and avian flu guidance (see Chap. 4). But, instead of simply incorporating challenges from existing guidance into the PREPARED Code, Vilma Lukaševičienė, who had undertaken the analysis of existing ethics guidance, compared the challenges she found with the articles in the draft PREPARED Code, a process that resulted in a small number of refinements, rather than new articles, including the following:

- "Quality controlled" was added to Article 1, which deals with the sharing of data about new infectious agents.
- "Health care responses" was modified to read "public health responses" in Article 14, which requires that such responses not be compromised by research.

10.4 Public Consultation

Public consultation was opened for seven weeks at the very end of the process, when the team had reached Version 12 of the PREPARED Code. Only one change was made in response to the public consultation, namely:

- The term "actors" in Article 12 was replaced with "all those involved in the research cycle".

10.5 The Final Draft

The handful of examples provided here give an indication of the level of consultation that led to Version 13. But they do not fully demonstrate how scrupulously every word in every article of the code was weighed. The most time-consuming element of the consultation process for the lead author was giving feedback, which she provided in writing to explain why suggestions might fall outside of the scope of the code. What was surprising was that the substance of the code changed little from Version 1 to Version 13, which is probably thanks to the comprehensive research foundation on which the first draft had been built.

11 Lessons from the PREPARED Code Approach

Writing the PREPARED Code was a massive undertaking: time-consuming, costly and complex, as summarised in this chapter. However, ensuring that a swift research response during pandemics is undertaken ethically is an aim worth investing in. By showing the depth of effort that went into the creation of the PREPARED Code, we hope we have helped demonstrate its credibility. As noted in Chap. 3, it is the behind-the-scenes *process* of code development that confers credibility (Messikomer and Cirka 2010).

Fairness and inclusivity guided the methodology in terms of evidence gathering in multiple languages, the inclusion of marginalised groups through sensitive and appropriate methods, the recruitment of global experts and stakeholders to the validation workshops, and the numerous rounds of consultation and feedback which ensured the refinements necessary to move from Version 1 to Version 13 of the code. Indeed, the PREPARED team engaged in dialogue with as many groups affected by the code as possible, and stimulated dialogue between these groups.

Through listening to the experiences and perspectives of a global sample of research stakeholders, the PREPARED team was able to develop a code that all stakeholders in the research process can reasonably accept. Furthermore, the risk-based approach, which focuses on real-world challenges, provides an important reality check. A major strength of the approach is that the ethical requirements are rooted in real-world risks drawn from diverse voices and experiences. It is worth noting that the risk-based approach avoids two potential problems: first, that something might be included in a new code simply because it was included in another ethics code, and second, that something might be included in a new code merely because guidance drafters *believe* it to be a problem.

The first problem – that of a requirement being transferred from another ethics code – can lead to ethics codes that are misaligned to their audience, because almost all early ethics codes were focused solely on biomedical research. New ethics codes might consequently be based on a codification of a particular type of research (i.e. biomedical research) that then imposes its ethics requirements on different types of research *inappropriately* (Yanow 2008; Schrag 2011).

The second problem – drafters including articles *they* think are necessary – can lead to a misalignment with the real problems researchers are likely to encounter. As the PREPARED team's approach of building all ethics guidance articles *solely* on real-world problems is unusual, an analogy can perhaps best illustrate this point.

The history of research ethics guidance shows that the vast majority of initiatives and guidance documents were driven from the standpoint of high-income countries (Resnik and Hofweber n.d.). At the same time, research has shown that research ethics committees from high-income countries can impose "remote paternalism" on researchers and research participants from lower-income countries (Schroeder et al. 2024: 32). One can therefore reasonably assume that there is a potential risk of misalignment between what ethics guidance drafters *think* are the challenges research ethics should seek to prevent and what these challenges really are, especially in a globalised world.

The values framework reflected in the TRUST and PREPARED Codes can be seen as a commitment to values that are commonly held globally and across cultures. While the methodology described above provided the space for other values to be identified, the final values of the PREPARED Code mirror the TRUST values of fairness, respect, care and honesty.

The approach is not, however, without limitations. The first draft of the code was developed after 18 months of research by a global consortium. This process was time-consuming and costly, requiring significant funding from the European Union, UK Research and Innovation and the Swiss State Secretariat for Education, Research and Innovation to implement. Such funds are not always available, which means that other groups might be unable to follow our approach on affordability grounds.

While we realise that not all initiatives will have access to the same resources, we hope to inspire groups tasked with developing professional codes of conduct in future to build their guidance on real-world problems and to be guided by the principles of fairness and inclusivity, making special efforts to involve the least privileged in decisions affecting our common futures.

References

1. ALLEA: The European Code of Conduct for Research Integrity, revised edn. All European Academies, Berlin (2017). https://www.allea.org/wp-content/uploads/2017/05/ALLEA-European-Code-of-Conduct-for-Research-Integrity-2017.pdf. Accessed 17 Feb 2025
2. Barrière, J., Frank, F., Besançon, L., et al.: Scientific integrity requires publishing rebuttals and retracting problematic papers. Stem Cell Rev. Rep. **19**, 568–572 (2023). https://doi.org/10.1007/s12015-022-10465-2
3. Bartmann, S., Erdmann, N., Haefker, M., et al.: Verstehendes Forschen in der Pandemie und anderen Ausnahmesituationen: praktische und methodologische Erkenntnisse der Rekonstruktiven Sozialen Arbeit (Understanding research in the pandemic and other exceptional situations: practical and methodological insights from reconstructive social work). Verlag Barbara Budrich, Leverkusen (2022)
4. Bassi, A., Arfin, S., Joshi, R., et al.: Challenges in operationalising clinical trials in India during the COVID-19 pandemic. Lancet Glob. Health **10**(3), e317–e319 (2022). https://doi.org/10.1016/s2214-109x(21)00546-5
5. Bermúdez, J.M.A., Maldonado, L.M.P.: La ética en la publicación científica en tiempos de COVID-19 (Ethics in scientific publishing in times of COVID-19). Revista de filosofía **38**(99), 225–240 (2021). https://doi.org/10.5281/zenodo.5644537
6. Cappelen, A., Falch, R., Sørensen, E., Tungodden, B.: Solidarity and fairness in times of crisis. J. Econ. Behav. Organ. **186**, 1–11 (2021). https://doi.org/10.1016/j.jebo.2021.03.017
7. Chatfield, K., Law, E.: 'I should do what?' Addressing research misconduct through values alignment. Res. Ethics **20**(2), 251–271 (2024). https://doi.org/10.1177/17470161231224481
8. Cohen, J.: Russia's approval of a COVID-19 vaccine is less than meets the press release. Science (2020). https://www.science.org/content/article/russia-s-approval-covid-19-vaccine-less-meets-press-release. Accessed 17 Jan 2025

9. Dadalto, L., Medeiros Royo, M., Silva Costa, B.: Bioética e integridade científica nas pesquisas clínicas sobre covid-19 (Bioethics and scientific integrity in clinical research on Covid-19). Revista Bioética **28**(3) (2020). https://doi.org/10.1590/1983-80422020283402
10. Dawson, A., Emanuel, E.J., Parker, M., et al.: Key ethical concepts and their application to COVID-19 research. Public Health Ethics **13**(2), 127–132 (2020). https://doi.org/10.1093/phe/phaa017
11. European Commission: Employment and social developments in Europe review: why social fairness and solidarity are more important than ever, 15 September 2020. https://ec.europa.eu/social/main.jsp?langId=en&catId=791&furtherNews=yes&newsId=9775. Accessed 27 Feb 2025
12. Fassin, Y.: Research on Covid-19: a disruptive phenomenon for bibliometrics. Scientometrics **126**, 5305–5319 (2021). https://doi.org/10.1007/s11192-021-03989-w
13. Ghooi, R.B.: Ethics committee meetings: online or face to face? Perspect. Clin. Res. **11**(3), 121 (2020). https://doi.org/10.4103/picr.PICR_97_20
14. Giorgini, V., Mecca, J., Gibson, C., et al.: Researcher perceptions of ethical guidelines and codes of conduct. Account. Res. **22**(3), 123–138 (2015). https://doi.org/10.1080/08989621.2014.955607
15. Humphries, F., Rourke, M., Berry, T., et al.: COVID-19 tests the limits of biodiversity laws in a health crisis: rethinking 'country of origin' for virus access and benefit-sharing. J. Law Med. **28**(3), 684–706 (2021). https://doi.org/10.2139/ssrn.3876009
16. Inserm: The VolREthics Initiative: healthy research volunteers and ethics (2022). https://www.inserm.fr/en/ethics/volrethics/. Accessed 28 Feb 2025
17. Jansen, Y.J., Foets, M.M., de Bont, A.A.: The contribution of qualitative research to the development of tailor-made community-based interventions in primary care: a review. Eur. J. Pub. Health **20**(2), 220–226 (2010). https://doi.org/10.1093/eurpub/ckp085
18. Jung, J., Kim, O.J.: Research ethics in COVID-19(SARS-CoV-2) vaccine study. Asia Paci. J. Health Law Ethics **13**(3), 1–25 (2020). https://doi.org/10.38046/apjhle.2020.13.3.001
19. Kadam, A.V., Patil, S., Sane, S., et al.: Challenges faced by ethics committee members in India during COVID-19 pandemic: a mixed-methods exploration. Indian J. Med. Res. **155**(5&6), 461–471 (2022). https://doi.org/10.4103/ijmr.ijmr_1095_22
20. Küçük, E.: The principle of solidarity and fairness in sharing responsibility: more than window dressing? Eur. Law J. **22**, 448–469 (2016). https://doi.org/10.1111/eulj.12185
21. Le Monde with AFP: Coronavirus: des spécialistes français s'excusent après leurs propos sur un test de vaccin en Afrique (Coronavirus: French specialists apologise after comments on vaccine test in Africa). Le Monde Afrique, 6 April 2020. https://www.lemonde.fr/afrique/article/2020/04/06/coronavirus-des-specialistes-francais-s-excusent-apres-leurs-propos-sur-un-test-de-vaccin-en-afrique_6035692_3212.html. Accessed 17 Jan 2025
22. Liu, D., Zhang, S., Zhou, J.: 新型冠状病毒肺炎临床研究的利益冲突管理(Conflict of interest management in clinical studies of novel coronavirus pneumonia). Chin. Med. Ethics **33**(10), 1185–1192 (2020)
23. Messikomer, C.M., Cirka, C.C.: Constructing a code of ethics: an experiential case of a national professional organization. J. Bus. Ethics **95**, 55–71 (2010). https://doi.org/10.1007/s10551-009-0347-y
24. Resnik, D.B., Hofweber, F.W.: Research ethics timeline (n.d.). https://www.niehs.nih.gov/research/resources/bioethics/timeline. Accessed 17 Feb 2025
25. Schrag, Z.M.: The case against ethics review in the social sciences. Res. Ethics **7**(4), 120–131 (2011). https://doi.org/10.1177/174701611100700402
26. Schroeder, D., et al.: The exclusion of vulnerable populations from research. In: Schroeder, D., et al. (eds.) Vulnerability Revisited. SpringerBriefs in Research and Innovation Governance, pp. 25–47. Springer, Cham (2024). https://doi.org/10.1007/978-3-031-57896-0_2

27. Schroeder, D., Chatfield, K., Singh, M., et al.: Equitable Research Partnerships: A Global Code of Conduct to Counter Ethics Dumping. Springer, Cham (2019). https://doi.org/10.1007/978-3-030-15745-6
28. Schroeder, D., Cook, J., Hirsch, F., et al.: Ethics Dumping: Case Studies from North-South Research Collaborations. SpringerBriefs in Research and Innovation Governance. Springer, Cham (2018). https://doi.org/10.1007/978-3-319-64731-9
29. Schveitzer, M.C., Thome, B.D.C.: Ética de investigación y asignación de recursos en tiempos de Covid-19 (Research ethics and resource allocation in times of Covid-19). Revista Bioética **29**(1), 21–26 (2021). https://doi.org/10.1590/1983-80422021291442
30. Schwandt, T.A.: Dictionary of Qualitative Inquiry, 2nd edn. Sage, Thousand Oaks (2001)
31. Sett, S., dos Santos, R.C., Prat, C., et al.: Access and benefit-sharing by the European virus archive in response to COVID-19. Lancet Microbe **3**(4), e316–e323 (2022). https://doi.org/10.1016/S2666-5247(21)00211-1
32. Tomson, G., Causevic, S., Ottersen, O.P., et al.: Solidarity and universal preparedness for health after covid-19. BMJ **372** (2021). https://doi.org/10.1136/bmj.n59
33. TRUST: The TRUST code: a global code of conduct for equitable research partnerships (2018). https://doi.org/10.48508/GCC/2018.05
34. Weijer, C.: COVID-19 human challenge trials and randomized controlled trials: lessons for the next pandemic. Res. Ethics **20**(4), 636–649 (2024). https://doi.org/10.1177/17470161231223594
35. Wright, S., O'Brien, B.C., Nimmon, L., et al.: Research design considerations. J. Grad. Med. Educ. **8**(1), 97–98 (2016). https://doi.org/10.4300/JGME-D-15-00566.1
36. Yanow, D., Schwartz-Shea, P.: Reforming institutional review board policy: issues in implementation and field research. PS Polit. Sci. Polit. **41**(3), 483–494 (2008). https://doi.org/10.1017/S1049096508080864
37. Zhang, Z., Li, M., Wang, J.: Subject protection in clinical trials of novel coronavirus pneumonia. Chin. Clin. Pharmacol. Ther. **25**(4), 421–425 (2020)

Open Access This chapter is licensed under the terms of the Creative Commons Attribution 4.0 International License (http://creativecommons.org/licenses/by/4.0/), which permits use, sharing, adaptation, distribution and reproduction in any medium or format, as long as you give appropriate credit to the original author(s) and the source, provide a link to the Creative Commons license and indicate if changes were made.

The images or other third party material in this chapter are included in the chapter's Creative Commons license, unless indicated otherwise in a credit line to the material. If material is not included in the chapter's Creative Commons license and your intended use is not permitted by statutory regulation or exceeds the permitted use, you will need to obtain permission directly from the copyright holder.

Implementation Support for the PREPARED Code

Carly Seedall[1(✉)], Doris Schroeder[2,3], Nearchos Paspallis[4,5], Thomas Nyirenda[6], and Michelle Singh[6]

[1] European Network of Research Ethics Committees, Bonn, Germany
seedall@eurecnet.eu
[2] School of Law, UCLan Cyprus, Larnaca, Cyprus
[3] Centre for Professional Ethics, UCLan, Preston, UK
[4] Interdisciplinary Centre for Law, Alternative and Innovative Methods, Larnaca, Cyprus
[5] UCLan Cyprus, Larnaca, Cyprus
[6] European and Developing Countries Clinical Trials Partnership, Cape Town, South Africa

Abstract. Research ethics and integrity codes lay the foundation for ethical research. However, stakeholders in research may struggle to move from reading codes to applying them, especially during times of crisis when there is increased uncertainty and risk. To bridge this gap, the PREPARED project team has developed adaptable tools to support implementation of the PREPARED Code for research ethics and research integrity during pandemics. These include training clips to accompany each code article, an experiential learning app, and how-to style guidelines for enhancing the resilience of stakeholder-specific processes. In this chapter, we summarise PREPARED's approach to code implementation, elaborate on each tool developed, and provide tips for future initiatives seeking to improve the practical application of ethics codes. We also present an example of how our tools were utilised during an African Vaccine Regulatory Forum (AVAREF) training session and provide a resource bank to support the integration of our materials into ethics training programmes.

Keywords: Research ethics and integrity guidance · research ethics and integrity training · research ethics and integrity case studies · the PREPARED app

1 Introduction

In the early 2020s, unprecedented collaboration between research bodies enabled the rapid development of COVID-19 vaccines and treatments. The research process was marked by urgency, taking place within multistakeholder networks (Xie et al. 2024). Moreover, research extended beyond the ivory tower of abstract theory. While policymakers and funders swiftly allocated resources, scientists spearheaded pandemic-related projects and healthcare workers, social workers and community leaders actively participated in research and took responsive actions on the front lines.

As a Nature (2021) editorial noted, the metaphor "standing on the shoulders of giants" widened in scope during the pandemic: "Today, such 'giants' are not only the investigators [...] but also every other participant in the research process. The future lies in standing on the shoulders of crowds."

Recognising the collective nature of this new research ecosystem, shaped by diverse contributors across disciplines and borders, the PREPARED Code seeks to ensure that research during pandemics can be accelerated without compromise of ethics and integrity values. But in the face of a crisis, can a global ethics code alone guarantee that research ethics and research integrity are upheld?

The world's largest ethics code library, the Illinois Institute of Technology's Codes of Ethics Collection, houses around 4,000 codes, including many with a focus on research ethics and research integrity (Illinois Tech n.d.; Sutrop et al. 2020). As research has become increasingly professionalised and institutionalised (Amsterdamski 1992), the development of ethics codes appears to have become standard practice.

However, concerns have emerged about whether ethics codes guide real-world ethical decision-making effectively. Although people value ethics codes, empirical research reveals a gap between their existence and practical use. For example, some working professionals admit to not using their codes or being ignorant of their content (Lere and Gaumnitz 2007). This finding extends to actors in research: in a small study, half of the researchers interviewed acknowledged that they relied solely on their institution's ethics codes and did not reference external ones at all (Schroeder et al. 2024). Thus, we cannot rely upon the existence of ethics codes alone to ensure adherence to research ethics and research integrity values. Bridging this gap demands an approach that translates ethical values and guidance into practical decision-making.

The effective implementation of ethics codes requires a clear understanding of the code guidance, which commonly takes the form of a number of *articles* (the specific rules, requirements or guidelines governing ethical behaviour). In this chapter, we explain how the PREPARED team set out to increase the impact of their code by creating a wide range of brief multimedia modules or texts, which we call training clips, that clarify the meaning of each article in the code and adapt it for real-world application.

Ethical dilemmas by their nature do not have straightforward solutions, which is why researchers are often trained to cultivate ethical reflexivity (von Unger 2021). In line with this approach, we explain how we developed an app, built on experiential learning techniques and incorporating a wide range of functionalities, to engage diverse learners and to catalyse ethical reflection.

We then explore how ethical decision-making can depend upon resilient research systems, highlighting the importance of procedural, stakeholder-specific how-to guidelines to complement ethics codes. To illustrate this, we present examples of the documents the PREPARED team developed to support the PREPARED Code, including guidelines on:

- fast-tracking ethics reviews to maintain fair and transparent yet accelerated desk reviews by journal editors
- prioritising research proposals in ethics committees without compromising review quality
- addressing the politicisation of science.

This chapter is a practical resource: as such, it provides tips in each subsection to guide readers in developing their own training to complement ethics codes. We also summarise the tools developed to support the PREPARED Code in a resource bank. In addition, to demonstrate the possible application of these tools, we explain how they were used to build a course for African regulators who assess clinical trials. The tools are summarised in Fig. 1.

Fig. 1. PREPARED Code implementation tools

2 Training Clips

Concrete examples make ethics code implementation more effective: they help actors understand how to apply guidance in practice, bridging the gap between abstract knowledge of ethics and sound decision-making in the real world (Schwarz 2004). However, if the intention is to keep the code globally relevant, concise and jargon-free, as with the PREPARED Code (see Chap. 3), these examples cannot be integrated into the code itself.

On the one hand, many codes are long and complex, which can induce code fatigue and render codes less user-friendly (Schwarz 2004; Lere and Gaumnitz 2007). On the other hand, overgeneralising in a code for the sake of brevity can lead to what has been called the "trap of analyticity" (Evers 2003, in Sutrop et al. 2020), where the need for broad consensus among heterogeneous actors results in vague provisions that lack clarity and conviction, ultimately reducing the ethics code's effectiveness. Thus, developers of ethics codes face a persistent challenge: balancing the need for wide-ranging applicability and inclusivity with the need for brevity and practical usefulness.

The global relevance of ethics codes is especially important during crises like pandemics. Crises require collaboration among diverse actors (Olsén et al. 2023), and a global framework provides a shared baseline of ethical values, ensuring coordinated and consistent responses under high-pressure conditions. For instance, during the COVID-19 pandemic, close collaboration among policymakers, funders, industry and ethics committees – particularly in regulatory processes and intellectual property negotiations – played a key role in enabling the development of vaccines in record time (Leisinger and Schroeder 2024).

Moreover, language plays an important role in the accessibility and usability of ethics codes. As Giorgini et al. (2015) argue, codes should be written in a way that allows them to be "readily encoded", necessitating clear and accessible wording. Global codes, however, often need to function across a range of linguistic and cultural contexts, which can introduce challenges. Tréguer-Felten (2017), for example, notes that some language used in global codes may be difficult to translate, while Adelstein and Clegg (2015) caution that "effusive and vague" language can impede understanding. To address these issues, the PREPARED team sought to provide additional context to mitigate linguistic and cultural misunderstandings related to the phrasing of the code's articles.

In response to the challenge of keeping a code practical and usable on a global scale while maintaining brevity and conciseness, the PREPARED team produced short training clips for each article of the PREPARED Code. These clips tie each article in the code directly to examples of the real-world risks that informed their creation. Consisting of short explanatory texts, videos and links to external sources, this material allows a deepened comprehension of the code by clarifying the meaning of the articles, grounding ethical values in real-world scenarios and ensuring accessibility without compromising depth. Where applicable, references were included to lend further credibility and offer pathways for extended learning.

An important technique to make training engaging is the use of professionally designed visual materials (Shabiralyani et al. 2015). Accordingly, the primary training material for the PREPARED Code relies heavily on visual content. Each of the 27 articles, along with select introductory sections, is elucidated through video clips. These clips are embedded where misunderstandings might arise, an approach we believe to be unique among ethics codes and one that we hope will significantly enhance understanding and uptake. In other words, the website (https://preparedcode.uclancyprus.ac.cy/) which presents the code is built in such a way that each article is accompanied by a short explanatory video.

For example, for Article 6 ("Research teams should share the additional responsibilities associated with a pandemic fairly among their members to avoid exacerbating existing inequalities"), a video was created highlighting how women researchers, disproportionately burdened by domestic and caregiving responsibilities in the workplace, experienced a decline in academic productivity during the pandemic compared to male researchers (Inguaggiato et al. 2024).

To clarify Article 1 ("Data and scientific insights about new infectious agents should be quality controlled and shared as swiftly as possible with the scientific community and other stakeholders, without prejudice to the sharer"), a senior South African law professor discusses how South African researchers did not hesitate to share data on the

Omicron variant of COVID-19, only for the country to face punitive travel restrictions in response.

The clips were produced in the following manner, summarised in Fig. 2 below:

1. Members of the PREPARED team with expertise in the relevant code article volunteered to create a training clip.
2. In collaboration with the lead author of the PREPARED Code, the most suitable format – animation, interview or video clip – was determined.
3. If an animation was selected (the most time-intensive option), the assigned experts drafted an initial script, aiming for a length of approximately one page.
4. The script was then reviewed for quality by other experts, as needed, and by the lead author. For clips incorporating video, images and graphics, a provisional voice-over was recorded to test clarity and effectiveness. This step helped eliminate overly long sentences and excessive jargon, as some script authors initially wrote in an academic style that was not suited to engaging visual content.
5. Once the script was finalised, the script authors selected visual materials from stock imagery and videos. Rather than having the designer make the initial selections, it was agreed that the experts would take the lead, with the designer assisting when needed.
6. The experts who wrote the script also selected appropriate stock music.
7. In most cases, the voice-over was provided by the colleagues who wrote the script, with the authors' own varied voices and accents regarded as more engaging than those of professional English-speaking voice artists.
8. The final clip was professionally edited and produced by an award-winning designer, who also identified and addressed any potential concerns.
9. After multiple iterations involving the script authors, the lead author and the designer, the final version of the clip was completed.

Developing effective training clips to clarify and contextualise an ethics code is key to bridging the gap between understanding and ethical decision-making. When creating such materials, future ethics projects could consider the following advice.

1. Ethics codes are more effective when accompanied by engaging, visually conceived training materials. Incorporate short videos, animations and explanatory clips to illuminate the guidance.
2. Instead of providing general training, rather target areas where comprehension challenges are likely to arise, especially due to linguistic or cultural differences.
3. Ensure that experts in the subject matter lead the creation of training materials to maintain accuracy and relevance. Where possible, encourage their collaboration with designers and communicators in order to enhance accessibility and avoid overly technical language.
4. While ethics codes often aim for global applicability, training materials should incorporate region-specific examples to enhance relatability. Ethics projects should address ethical challenges in different cultural and regulatory contexts to promote greater engagement.
5. Allow time for repeated rounds of review of the training clips, including quality control by experts and practical testing, to eliminate jargon and ensure clarity.

Fig. 2. Steps in the development of training clips

3 The PREPARED App

Ethics training has two main aims. The first substantive aim focuses on increasing understanding and raising awareness of specific issues (Montgomery and Walker 2012). The second seeks to enhance engagement in deep ethical reflection and the development of ethical competency (Andersson et al. 2022). To address both aims, the PREPARED team developed a mobile app offering full training courses with information on ethical and integrity-related aspects and presenting case study dilemmas designed to encourage ethical reflection.

The effectiveness of case studies, which allow learners to apply theoretical decision-making frameworks to real-world situations, stems from their ability to elicit active participation from learners (Escartín et al. 2015). For this reason, the PREPARED team developed a diverse set of case studies based on ethical dilemmas that research actors face during crises.

For training to be effective, it must be both accessible and engaging. As 69% of the world's population owns a smartphone (Laricchia 2024), the team opted to develop a mobile app to present these training elements. Beyond basic text and images, engagement is fostered through interactive elements in the app, like polls, multiple-choice questions, mini-games, simulated dialogues and professionally produced film clips and animations to encourage active learning. For example, when presented with an ethical dilemma, users participate in a poll and can later view aggregated responses from other learners, prompting reflection on their own decisions.

To accommodate the limitations of mobile devices, each case study is structured into pages corresponding to smaller steps that learners complete progressively. This allows users to proceed through the material at their own pace, either finishing a case study in one session or pausing and resuming as needed. Figure 3 shows several screenshots from the app.

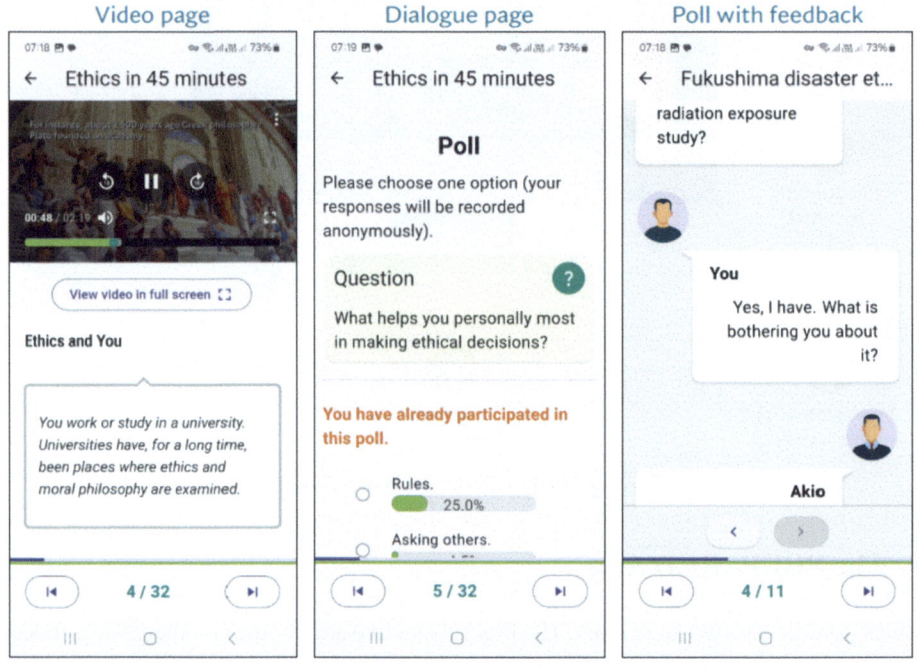

Fig. 3. Screenshots from the app

A key interactive feature is the dialogue simulation, designed to resemble a text-based conversation. Learners navigate a step-by-step discussion, taking part in a simulated exchange where two individuals debate different aspects of an ethical dilemma. In addition, mini-games, such as a decision-making exercise, prompt learners to assess and categorise factors related to the case – such as pros and cons – while receiving feedback on their choices.

With its accessible and interactive design, designed primarily for mobile devices, the PREPARED app is open to a broad range of stakeholders, including researchers and students. Its clear, jargon-free approach also makes it suitable for the public, including those not directly involved in research.

At the time of writing, the PREPARED app contained 27 professionally designed video clips, ranging in length from two minutes to four minutes 45 s. It currently includes two short courses and five case studies, which are described in Table 1.

Table 1. PREPARED app courses and case studies

Type	Topic	Overview
Short course	Ethics in 45 min	The role of rules, virtues, values and ethics codes in ethical decision-making, and an ethical dilemma applying the values of fairness and respect
	The TRUST Code in 45 min	Inequities in global research, such as ethics dumping and helicopter research, and how they can be tackled through the TRUST Code: A Global Code of Conduct for Equitable Research Partnerships
Case dilemma	Are SARS-CoV-2 human challenge studies ethical?	Pros and cons of human challenge studies to be investigated by the learner before providing their own view
	AI ethics and research during crisis	AI research that is meant to contribute to achieving the UN's Sustainable Development Goals, but can have severe negative implications for the worst off
	Ethical challenges for research ethics committees during COVID-19	A fictional dilemma about a psychology project seeking approval by a research ethics committee
	Scientific collaboration during war	The positions taken around the world on whether to collaborate with Russian institutions following the Ukraine war, and finding one's own position among the possibilities
	Navigating ethical challenges following the nuclear accident in Fukushima	A data and informed consent dilemma that examines the borderline between research and public health crisis management

In summary, we would advise future ethics projects to consider the following:

1. Ethical decision-making is best learned through context. Develop diverse case studies that reflect real-world dilemmas. These can accompany informative course materials to enhance understanding of relevant ethics and integrity matters.

2. Deliver ethics training via a mobile app to promote broad accessibility. When planning an app, design content for smaller screens and be aware of the limitations of mobile functionalities.
 3. Include interactive features such as polls, multiple-choice questions, simulated dialogues and mini-games.
 4. Combine text with professionally designed videos, animations and audio elements to increase learners' retention.

4 How-To Guidelines

Ethics codes can establish foundational values and principles while also offering stakeholder-specific guidance. However, during crises, the practical implementation of this guidance can be challenging, as research systems rely on specific processes that may be disrupted. To address this, the PREPARED team developed how-to guidelines tailored to the research governance systems of groups like research ethics committees, publishers and research integrity officers. These are intended to pinpoint steps to improve these systems during "normal times", thereby strengthening the resilience of the research ecosystem.

Though resilience has many definitions, they generally emphasise minimising the negative impact that a crisis has on a "system's performance" (Hosseini et al. 2016). Additionally, resilience must be proactively built through coordination within these systems during "normal times" rather than only in response to crises (Reiss et al. 2024). To strengthen resilience in existing research systems during periods of non-crisis, the PREPARED team developed practical how-to guidelines tailored to different research stakeholders and designed for fluid integration into existing processes within the research ecosystem.

Importantly, one set of guidelines also addresses the proliferation of ethics guidance (see Chap. 3) in an unusual way. Generally, ethics codes do not "work" with each other. While there may be cross-references (for instance, the PREPARED Code cross-references to the Declaration of Helsinki), there is generally no deeper engagement. To increase the usefulness of the PREPARED Code, the team therefore took an innovative step at the recommendation of the European & Developing Countries Clinical Trials Partnership (EDCTP).

In 2020, the UK Collaborative on Development Research (UKCDR) and the Global Research Collaboration for Infectious Disease Preparedness (GloPID-R), a funder network focused on new or re-emerging infectious diseases, issued a set of seven principles to encourage high-quality, ethical research during epidemics and pandemics (Norton et al. 2020). These are:

 1. alignment to global research agendas and locally identified priorities
 2. research capacity for rapid research
 3. supporting equitable, inclusive interdisciplinary and cross-sectoral partnerships
 4. open science and data sharing
 5. protection from harm
 6. appropriate ethical consideration
 7. collaboration and learning through enhanced coordination.

The seven principles build on best-practice guidance generated by the earlier work of UKCDR, GloPID-R, the World Health Organization (WHO), the European Commission and others. They provide a basis for guiding both funder and researcher expectations for COVID-19 and for future epidemics and pandemics. The principles are globally relevant and of particular importance for research in low-resource settings (Norton et al. 2020).

As the EDCTP is part of this initiative *and* part of the PREPARED team, it encouraged the team to examine possible links between the two. The PREPARED Code was found to be an ideal companion to help operationalise the seven principles set out above. A guideline explaining how this can be done is due to be published in June 2025, and should reduce the proliferation of ethics codes by encouraging innovative collaborations.

Schwartz (2004) shows that people are more likely to remember parts of a code that relate to their everyday tasks. When codes are disconnected from processes within these systems, users may view them as irrelevant (Marnburg 2000; Salvioni et al. 2015). Effective ethics codes must therefore be easily "translated into practical action" and be embedded within systems (Lindner 2014). In an increasingly formalised research environment (Shaw et al. 2005), research is advanced through systems like funding pipelines, ethics reviews and publishing protocols that must be targeted through tailored guidelines to make ethics codes effective.

To identify the stakeholder groups requiring specific guidance, the PREPARED team leveraged stakeholder engagement platforms. These included researchers, funders, non-governmental organisations, publishers and editors, industry representatives and governance actors such as research ethics committees and research integrity officers. Insights into their needs were gathered through surveys, focus groups and stakeholder-specific literature reviews (Seedall and Tambornino 2022).

Throughout the development of the PREPARED Code, which included gathering empirical evidence and facilitating consultations with stakeholder groups, the team analysed existing research systems to understand how ethical values and principles were applied in practice and to identify operational gaps where additional support may be needed before the advent of the next crisis. Concise, jargon-free guidance was subsequently developed to help stakeholders implement research governance processes aligned with the values of the PREPARED Code.

For example, during the COVID-19 pandemic, research ethics committees faced overwhelming pressure to process a surge of COVID-19-related research proposals (Reyes 2020). A survey of European research ethics committees (Seedall and Tambornino 2022), scoping reviews (e.g. Seedall and Tambornino 2024), an analysis of existing ethics codes and validation workshops conducted with research ethics committee members revealed a key challenge: research ethics committees were inundated with research proposals related to COVID-19, many of which were of low quality. Despite this, they were tasked with prioritising studies that addressed urgent societal needs. To navigate this challenge, committees required guidance on streamlining their review processes while maintaining standards for ethical research (Tamariz et al. 2021).

To respond to these challenges, a set of recommendations for expediting ethics review during crises was developed (Kornioti et al. 2024). Kornioti et al. (2024) identify seven key obstacles that research ethics committees encountered during the COVID-19 pandemic. Their report offers practical strategies for fast-tracking research protocols. Each challenge is paired with real-life examples of good practice implemented by research ethics committees during the pandemic, making the guidance practicable. The recommendations focus primarily on strengthening institutional processes during stable periods to prepare better for future crises. For example, committees and research institutions are advised to adopt remote workflows, establish systems of mutual recognition and implement sustainable funding and compensation models.

In addition, prioritisation guidelines were developed for research ethics committees, though these have not yet been published at the time of writing. Based on the results of a survey of more than 320 research ethics committees from over 80 countries, this document will guide research ethics committees when deciding how systems of prioritisation could be changed, the criteria upon which these decisions should be based, and the potential implications of such decisions. These guidelines will enhance resilience by enabling committees to establish and justify prioritisation decisions in advance, and helping them to efficiently manage high volumes of protocol submissions during crises.

Downstream – in the research dissemination process – academic journals also faced a very high increase in submissions. Throughout the project, publishers and editors informed the team that journals, already overstretched, had been inundated with manuscripts during the COVID-19 pandemic. Many lacked streamlined processes for the swift and fair evaluation of submissions (Seedall and Tambornino 2022), which made it difficult for publishers and editors to disseminate research findings promptly.

To address this, guidance for editors and publishers was developed, drawing from a targeted literature review and consultations with stakeholders (Chatfield 2024). This guidance suggests measures for streamlining the initial review stage, enabling editorial teams to assess manuscripts transparently and efficiently. The recommendations aim to support the swift publication of research while maintaining fairness and quality in editorial desk assessments. These recommendations can improve resilience as they help to ensure that editorial systems are equipped with transparent processes to handle surges in submissions.

Lastly, guidance is being developed to address harassment and the politicisation of science, though it has not yet been published. Validation workshops conducted during the project, as well as a survey carried out by the Finnish Committee for Public Information (TJNK 2024), revealed that researchers are increasingly concerned about the risks of speaking publicly about research, particularly due to harassment on social media. During the COVID-19 pandemic, some early-career researchers avoided studying controversial topics in order to escape potential harassment, which amounted to self-censorship. While social media was a major source of harassment, some incidents also originated within the research community itself, including conflicts related to workplace dynamics and internal disagreements.

In response, guidelines developed in a collaboration between the PREPARED team and TENK, the Finnish National Board on Research Integrity, will outline steps for funders, research institutions and policymakers to develop mechanisms to monitor the environment surrounding their researchers. These guidelines will support resilience, creating systems that address harassment and conflicts before the next crisis to reduce the long-term impact on researchers and the broader research community.

Developing effective how-to guidelines for implementing ethics codes requires an approach that is both systematic and collaborative. From our experience, this process can be conducted in alignment with the following steps:

1. First, identify and liaise with the stakeholder groups for whom the guidelines will be written. These groups may include researchers, research ethics committees, publishers, funders and industry representatives. Establishing these connections ensures that the guidelines address real-world challenges and reflect diverse perspectives.
2. Next, communicate with individual stakeholder groups to uncover challenges within their governance systems which may be exacerbated by a crisis. During the development of an ethics code, gaps and operational questions often emerge. Engaging with stakeholders through surveys, focus groups or consultations can highlight specific pain points and areas where procedural support is needed to complement an ethics code.
3. Once challenges have been identified, draft jargon-free procedural documents tailored to each stakeholder group. These documents should provide practicable steps that stakeholders can take and integrate into their existing systems in "normal times". Avoid technical language to ensure that the guidelines are accessible to a wide audience.
4. Validate these documents through stakeholder input and refine them. Sharing drafts with stakeholders and incorporating their feedback not only ensures that the guidelines are applicable, it also fosters stakeholder buy-in and increases the likelihood of successful adoption.
5. Finally, launch guidelines in high-profile venues to relevant stakeholders. For instance, the PREPARED Code and the harassment guidelines will be launched at an event hosted by UNESCO, at their headquarters in Paris, in June 2025.

5 Implementation Example: AVAREF Training of African Regulators

In this section, using examples from training sessions that were developed for the African Vaccine Regulatory Forum (AVAREF), we illustrate how the PREPARED materials can be effectively adapted to regional settings.

AVAREF, managed by the WHO Regional Office for Africa (WHO AFRO) and supported by the EDCTP, deployed facilitators to regional training sessions to equip African regulators with comprehensive skills in clinical trials assessment and oversight. Training was conducted in the three official working languages of WHO AFRO, English, French and Portuguese. In 2024, this training was conducted in Namibia and Senegal.

The training sessions aimed to equip selected nominees from regulatory authorities in African regional economic communities with the skills to assess nonclinical data, clinical trial data, biostatistics data and clinical quality data, as well as to improve knowledge regarding emergency use authorisation (in accordance with World Health Assembly resolution WHA75.8; see Fig. 4).

The AVAREF training incorporated both TRUST and PREPARED materials, with a targeted session on equitable research partnerships and another on research during pandemics. These sessions included videos on the TRUST Code as well as case studies

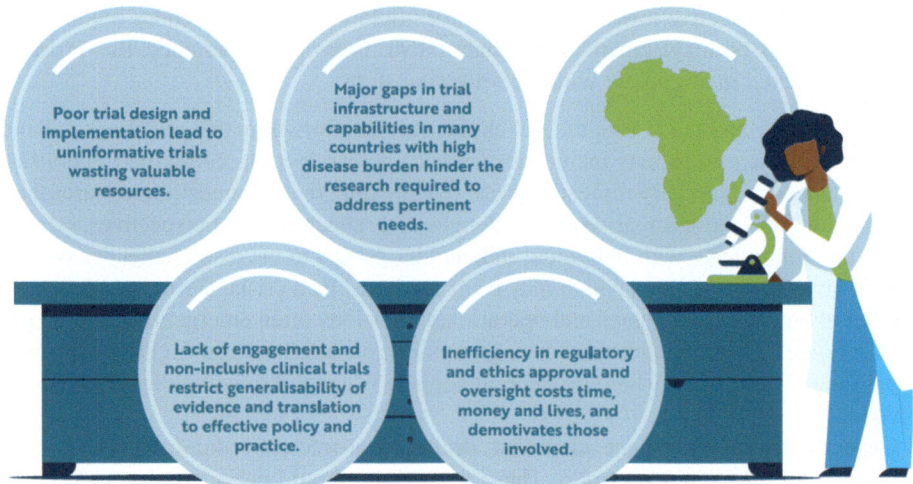

Fig. 4. Background to WHA75.8: Barriers in clinical trials that put public health at risk

from the PREPARED app. Participants were also invited to provide feedback on a draft of the PREPARED Code (see Chap. 5).

Trainers emphasised the need to introduce the foundations of both ethics codes, including their values-based approach, before discussing specific articles. This was achieved using training clips and the TRUST Code training programme available in the PREPARED app. Trainers also found the translated versions of the codes useful for participants whose first language was not English.

When using case studies, trainers observed high levels of engagement but noted that participants needed an overview of the activity and a clear explanation of its purpose – specifically, why working with ethical dilemmas was useful and how ethical reflection informed decision-making. Trainers also found that discussing case studies in smaller groups before sharing insights with the larger group was a more effective approach than purely whole-group discussions.

The training revealed the need for a wider range of case studies. Participants preferred discussing cases relevant to their region, reinforcing the need to adapt training materials to different global contexts.

Finally, format was key to making the training effective. Trainers found that video clips, for example, maintained trainee attention better than traditional slide presentations.

6 Resource Bank

Where multiple complementary resources exist, some training materials may lack discoverability. For that reason, we have summarised our training materials below in the form of a resource bank. This resource bank includes the training clips, mobile app and how-to guidelines described above, with a brief description of each training resource and where to find it.

6.1 The PREPARED Code

The PREPARED Code: A Global Code of Conduct for Research During Pandemics
The PREPARED Code targets researchers, research ethics committees and research integrity officers and is reproduced in Chap. 2. The code was written in English and has been translated into 11 languages: Arabic, simplified Chinese, Finnish, French, German, Greek, Italian, Korean, Lithuanian, Portuguese, Spanish and Swahili. Further information about the code and the downloadable translations are available online here: https://prepared-project.eu/prepared-code/

The TRUST and PREPARED Values
The PREPARED Code is underpinned by the values of fairness, respect, care and honesty, the same values that underpin the TRUST Code. This video explains how the fundamental values of the TRUST Code also apply to research challenges in global crises, such as the COVID-19 pandemic: https://youtu.be/LEUXu-ZyhYg.

6.2 PREPARED Code Training Resources

Article Training Clips
The home of the PREPARED Code contains 27 concise clips linked to each article of the code. These clips aim to contextualise and clarify the articles of the code. They include videos, filmed interviews and references to additional publications and guidelines developed within the project, and are available here: https://preparedcode.uclancyprus.ac.cy.

The PREPARED App
The PREPARED app provides a digital platform to complement training in research ethics and research integrity decision-making during global crises. The app presents research ethics training that is engaging, interactive and conveniently packaged so that it is accessible using a smartphone (Android and iOS). Functionalities include polls, sorting buckets, quizzes, professionally designed animations, interview clips, guided dialogues and audio clips. Free download of the app is available on the Apple App Store and Google Play Store: https://prepared-project.eu/app/

6.3 PREPARED How-To Guidelines

Recommendations for Expediting Ethics Review during Times of Crisis
This guidance sets out seven major challenges that research ethics committees experienced during the COVID-19 pandemic. It provides recommendations for effective fast-tracking of study protocols and a good-practice example for each challenge. Available here: https://prepared-project.eu/fast-track-guidance/

Guidance for Fair and Fast Desk Assessment of Submitted Manuscripts during Times of Crisis
This guidance for editors and publishers covers the process for identifying manuscripts that meet the threshold criteria for peer review and the criteria against which submissions can be assessed fairly. Available here: https://prepared-project.eu/fast-track-guidance/

6.4 TRUST Code Training Resources

The TRUST Code in 45 min

This is a short course to explain the development of the value-based TRUST Code: A Global Code of Conduct for Equitable Research Partnerships.

The training is available in two versions: first, by free download of the PREPARED app, as mentioned earlier: https://prepared-project.eu/app/ and second, as a stand-alone web-based training resource: https://trustcodetraining.uclancyprus.ac.cy.

Short Video Clip about the TRUST Code

For those not using the 45-minute course on the TRUST Code, we recommend the screening of a short clip about the TRUST Code. Available online here: https://youtu.be/3nRFWNmx1Y4.

6.5 Research Ethics and Research Integrity Training Materials

Ethics in 45 min

This short course, developed by the University of Central Lancashire for students and academics, provides a concise and motivational introduction to ethics and ethical decision-making. It connects ethical thinking across ages and continents; distinguishes between rules, values and virtues; explains different types of ethics codes; and tests insights into a moral dilemma. The short course is available via the PREPARED App, which can be downloaded free from the Apple App Store and the Google Play Store: https://prepared-project.eu/app/

Training Videos

Partners in the PREPARED consortium have created short videos to slot into training programmes. These include the following:

- Who benefits in research?
- Benefit-sharing
- AI ethics in five minutes
- Ethics dumping
- Ethical controversies around human challenge studies
- Healthy volunteers and human challenge studies
- AI ethics and helicopter research
- AI ethics and the Sustainable Development Goals
- Scientific collaboration during war
- Lockdown and the experience of Nairobi sex workers

The videos are available at https://www.youtube.com/@trustandprepared1000/videos and https://prepared-project.eu/free-training-materials/

7 Conclusion

During the COVID-19 pandemic, research was shouldered by crowds rather than giants (Nature 2021). Hence, ethics guidance needs to be relevant to a broad range of stakeholders. At the same time, we argue that global ethics codes alone are not enough to ensure ethical decision-making. This raises the question: how can a global ethics code be both effective and relevant to a growing plurality of actors in the research ecosystem?

The PREPARED project answer to this lies in the development of a range of resources that are intended to support understanding and implementation of the PREPARED Code. Above all, the materials aim to promote ethical reflection, encouraging those involved in research to go beyond theoretical understanding and actively consider the dilemmas they may face during crises. This emphasis on ethical reflexivity is particularly evident in the case studies developed for the PREPARED app.

The PREPARED Code is designed for a global audience. Alongside its sister code, the TRUST Code – which has been adopted by institutions worldwide – the PREPARED Code has been developed by a diverse and engaged international team. To keep the code succinct while ensuring its accessibility across cultural and linguistic contexts, training clips were developed that clarify and contextualise each article in the code.

Recognising that a resilient research ecosystem relies on the adaptation of existing research processes, stakeholder-specific how-to guidelines were developed that speak to the procedures already being followed by research ethics committees, publishers and editors, and research-performing institutions.

Formats that enhance engagement, such as video clips, polls and interactive dialogues, were prioritised. To ensure accessibility, many of the training materials are hosted on a mobile app, the PREPARED app, making them available to a global audience.

In summary, we strongly encourage future ethics projects to support implementation of their codes and guidance with tools that enhance ethical reflection, global applicability, accessibility and system resilience, and include engaging formats.

As crises continue to reshape the global research landscape, the challenge is no longer just to develop ethics codes: we must ensure that they are both understandable to diverse actors and applied appropriately to guide ethical decision-making.

References

Adelstein, J., Clegg, S.: Code of ethics: a stratified vehicle for compliance. J. Bus. Ethics **138**(1), 53–66 (2015). https://doi.org/10.1007/s10551-015-2581-9

Amsterdamski, S.: The institutionalization and professionalization of scientific research. In: Amsterdamski, S. (ed.) Between History and Method. Boston Studies in the Philosophy of Science, vo.1 145, pp. 65–78. Springer, Dordrecht (1992) . https://doi.org/10.1007/978-94-011-2706-6_5

Andersson, H., Svensson, A., Frank, C., et al.: Ethics education to support ethical competence learning in healthcare: an integrative systematic review. BMC Med. Ethics **23**(1), 29 (2022). https://doi.org/10.1186/s12910-022-00766-z

Chatfield, K.: Guidance for fair and fast desk assessment of submitted manuscripts during times of crisis. A report for PREPARED (2024). https://prepared-project.eu/wp-content/uploads/2024/03/Guidance-for-fair-and-fast-desk-assessment.pdf. Accessed 14 Feb 2025

Escartín, J., Saldaña, O., Martín-Peña, J., et al.: The impact of writing case studies: benefits for students' success and well-being. Procedia Soc. Behav. Sci. **196**, 47–51 (2015). https://doi.org/10.1016/j.sbspro.2015.07.009

Evers, K.: Codes of conduct: standards for ethics in research. Office for Official Publications of the European Communities, Luxembourg (2003)

Giorgini, V., Mecca, J.T., Gibson, C., et al.: Researcher perceptions of ethical guidelines and codes of conduct. Account. Res. **22**(3), 123–138 (2015). https://doi.org/10.1080/08989621.2014.955607

Hosseini, S., Barker, K., Ramirez-Marquez, J.E.: A review of definitions and measures of system resilience. Reliab. Eng. Syst. Saf. **145**, 47–61 (2016). https://doi.org/10.1016/j.ress.2015.08.006

https://prepared-project.eu/wp-content/uploads/2024/03/Recommendations-for-expediting-ethics-review-during-times-of-crisis.pdf. Accessed 14 Feb 2025

Illinois Tech: Center for the Study of Ethics in the Professions: About. Illinois Institute of Technology, Chicago IL (n.d.). https://www.iit.edu/center-ethics/about. Accessed 19 Dec 2024

Inguaggiato, G., Pallise Perello, C., Verdonk, P., et al.: The experience of women researchers during the COVID-19 pandemic: a scoping review. Res. Ethics **20**(4), 780–811 (2024). https://doi.org/10.1177/17470161241231268

Kornioti, N., Seedall, C., Kalaitzaki, K., Marcou, A.: Recommendations for expediting ethics review during times of crisis. A report for PREPARED (2024)

Laricchia, F.: Global smartphone penetration as share of population from 2016 to 2023. Statista (2024). https://www.statista.com/statistics/203734/global-smartphone-penetration-per-capita-since-2005. Accessed 19 Dec 2024

Leisinger, K., Schroeder, D.: Project lightspeed: a case study in research ethics and accelerated vaccine development. Res. Ethics **20**(4), 847–856 (2024). https://doi.org/10.1177/17470161241251597

Lere, J.C., Gaumnitz, B.R.: Changing behavior by improving codes of ethics. Am. J. Bus. **22**(2), 7–18 (2007). https://doi.org/10.1108/19355181200700006

Lindner, S.: Implementing codes of conduct in public institutions. Transparency International (2014). https://knowledgehub.transparency.org/helpdesk/implementing-codes-of-conduct-in-public-institutions. Accessed 14 Feb 2025

Marnburg, E.: The behavioural effects of corporate ethical codes: empirical findings and discussion. Bus. Ethics Eur. Rev. **9**(3), 200–210 (2000). https://doi.org/10.1111/1467-8608.00191

Montgomery, D., Walker, M.: Enhancing ethical awareness. Gift. Child Today **35**(2), 95–101 (2012). https://doi.org/10.1177/1076217511436092

Nature: Research collaborations bring big rewards: the world needs more (editorial). Nature **594**(7863), 301–302 (2021). https://doi.org/10.1038/d41586-021-01581-z

Norton, A., Mphahlele, J., Yazdanpanah, Y., et al.: Strengthening the global effort on COVID-19 research. Lancet **396**(10248), 375 (2020). https://doi.org/10.1016/S0140-6736(20)31598-1

Olsén, M., Oskarsson, P.A., Hallberg, N., et al.: Exploring collaborative crisis management: a model of essential capabilities. Saf. Sci. **162**, 106092 (2023). https://doi.org/10.1016/j.ssci.2023.106092

Reiss, M., Kraus, M., Riedel, M., Czypionka, T.: What makes health systems resilient? An analytical framework drawing on European learnings from the COVID-19 pandemic based on a multitiered approach. BMJ Public Health **2**(1), e000378 (2024). https://doi.org/10.1136/bmjph-2023-000378

Reyes, M.: Research in the time of COVID-19: challenges of research ethics committees. J. ASEAN Fed. Endocr. Soc. **35**(1), 29–32 (2020). https://doi.org/10.15605/jafes.035.01.07

Salvioni, D.M., Astori, R., Cassano, R.: Corporate sustainability and ethical codes effectiveness. SSRN Electron. J. (2015). https://doi.org/10.2139/ssrn.2577393

Schroeder, D., Chatfield, K., Chennells, R., et al.: Vulnerability Revisited: Leaving No One Behind in Research. Springer, (2024) https://doi.org/10.1007/978-3-031-57896-0

Schwartz, M.S.: Effective corporate codes of ethics: perceptions of code users. J. Bus. Ethics **55**(4), 321–341 (2004). https://doi.org/10.1007/s10551-004-2169-2

Seedall, C., Tambornino, L.: Research ethics and integrity in times of global crisis: stakeholder needs. A report for PREPARED (2022). https://prepared-project.eu/wp-content/uploads/2023/12/NEW-PREPARED-Stakeholder-needs-Milestone-Final.pdf. Accessed 14 Feb 2025

Seedall, C., Tambornino, L.: Research ethics and integrity in the Dach region during the COVID-19 pandemic: balancing risks and benefits under pressure. Res. Ethics **20**(4), 650–668 (2024). https://doi.org/10.1177/17470161241229207

Shabiralyani, G., Hasan, K., Hamad, N., Iqbal, N.: Impact of visual aids in enhancing the learning process case research: District Dera Ghazi Khan. J. Educ. Pract. **6**(19), 226–233 (2015). https://files.eric.ed.gov/fulltext/EJ1079541.pdf. Accessed 14 Feb 2025

Shaw, S., Boynton, P.M., Greenhalgh, T.: Research governance: where did it come from, what does it mean? J. R. Soc. Med. **98**(11), 496–502 (2005). https://doi.org/10.1258/jrsm.98.11.496

Stevens, B.: Corporate ethical codes: effective instruments for influencing behavior. J. Bus. Ethics **78**(4), 601–609 (2007). https://doi.org/10.1007/s10551-007-9370-z

Sutrop, M., Parder, M.L., Juurik, M.: Research ethics codes and guidelines. In: Iphofen, R. (eds.) Handbook of Research Ethics and Scientific Integrity, pp. 1–23. Springer, Cham (2020). https://doi.org/10.1007/978-3-319-76040-7_2-1

Tamariz, L., Hendler, F.J., Wells, J.M., et al.: A call for better, not faster, research ethics committee reviews in the Covid-19 era. Ethics Hum. Res. **43**(5), 42–44 (2021). https://doi.org/10.1002/eahr.500104

TJNK: Häiritsevän palautteen monet muodot 2023 (Many forms of disturbing feedback 2023). Tiedonjulkistamisen neuvottelukunta, Helsinki (2024). https://tjnk.fi/fi/vaikuttaminen/tutkijan-sananvapaus/hairitsevan-palautteen-monet-muodot-2023. Accessed 29 Jan 2025

Tréguer-Felten, G.: The role of translation in the cross-cultural transferability of corporate codes of conduct. Int. J. Cross Cult. Manag. **17**(1), 137–149 (2017). https://doi.org/10.1177/1470595817694659

von Unger, H.: Ethical reflexivity as research practice. Hist. Soc. Res. Historische Sozialforschung **46**(2), 186–204 (2021). https://doi.org/10.12759/hsr.46.2021.2.186-204

Xie, H., Guo, M., Yang, Y.: Exploring the processes and mechanisms by which nonprofit organizations orchestrate global innovation networks: a case study of the COVAX program. Heliyon **10**(5) (2024). https://doi.org/10.1016/j.heliyon.2024.e27098

Open Access This chapter is licensed under the terms of the Creative Commons Attribution 4.0 International License (http://creativecommons.org/licenses/by/4.0/), which permits use, sharing, adaptation, distribution and reproduction in any medium or format, as long as you give appropriate credit to the original author(s) and the source, provide a link to the Creative Commons license and indicate if changes were made.

The images or other third party material in this chapter are included in the chapter's Creative Commons license, unless indicated otherwise in a credit line to the material. If material is not included in the chapter's Creative Commons license and your intended use is not permitted by statutory regulation or exceeds the permitted use, you will need to obtain permission directly from the copyright holder.

Learning from the PREPARED Experience: Recommendations for Enhancing the Effectiveness and Credibility of New Ethics Codes

Kate Chatfield[1(✉)] and Michelle Singh[2]

[1] Centre for Professional Ethics, UCLan, Preston, UK
kchatfield@uclan.ac.uk
[2] European and Developing Countries Clinical Trials Partnership Association, Cape Town, South Africa

Abstract. As the risks of ethics and integrity breaches are higher during times of crisis, guidance that enables accelerated research without violating ethics values is essential. This chapter draws upon the lessons learned from a broad range of activities underpinning the development of the PREPARED Code to make recommendations for future developers of ethics codes. Recommendations take the form of key ingredients to help future developers enhance the effectiveness and credibility of ethics codes: building the code on real world risks, aligned with moral values, through transparent and inclusive development processes and with implementation support.

Keywords: Recommendations · ethics code development · inclusivity · moral values · risk-based · accessibility

1 Introduction

The chapters in this book guide the reader through the process of developing the PRE-PARED Code. In this final chapter, we draw from the PREPARED experience to present recommendations which we hope will serve as a valuable resource for future developers of ethics codes. These recommendations take the form of six key ingredients that we put forward to help enhance the effectiveness and credibility of new ethics codes (see Fig. 1).

One might question our authority to make recommendations, given that, at the time of writing, the PREPARED Code has not yet been implemented. How can we assume its effectiveness and credibility without real-world testing? The truth is, we cannot be certain. However, the TRUST Code (TRUST 2018), which was developed using the same methodological approach, has proven to be highly impactful (Chatfield and Law 2024). And in developing the PREPARED Code, the team drew upon the lessons learned from the TRUST experience, refining the process and adapting it to the context of a pandemic.

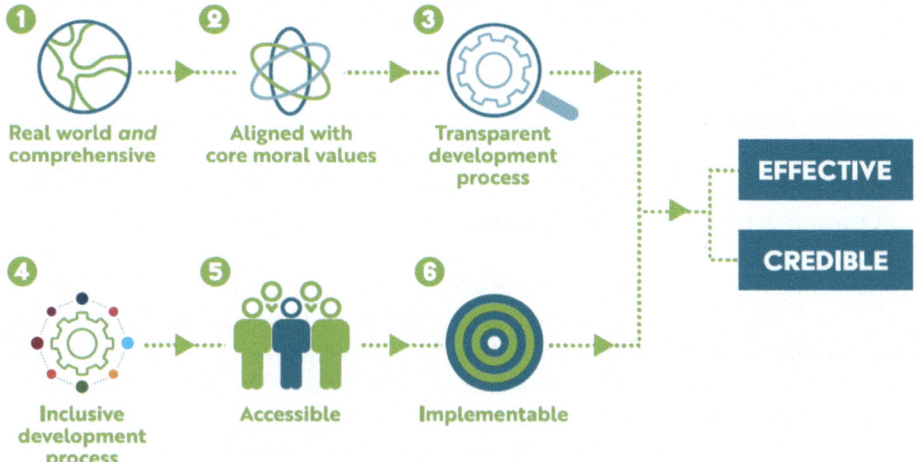

Fig. 1. Six key ingredients for the development of a new ethics code

Thus, we offer these recommendations in the spirit of sharing, hoping that insights from our experience might help to support future developments.

Nevertheless, what makes the PREPARED Code approach unique is the *combination* of these six ingredients to enhance effectiveness and credibility. This is especially important when a new ethics code is being developed for unfamiliar contexts – contexts in which, as was the case for an ethics code for research during pandemics, there is no existing, time-tested code.

Inspired by the TRUST and PREPARED experiences, the following sections explain the recommended six key ingredients for ethics code development.

2 Real World and Comprehensive

A fundamental first step in the development of any new code of conduct is deciding what ethical issues or risks need to be addressed. As explained in Chap. 3, there are various ways of doing this, for instance by drawing upon existing codes or the experiences and knowledge of experts and code drafters. However, these methods might lead drafters to include issues simply because they appear in existing ethics codes, or because the guidance drafters or experts *assume* them to be problems. The concern here is that challenges identified in this way might not reflect what happens in the real world, or might not capture the full extent of the challenges.

Alternatively, one can take a risk-based approach to identifying what needs to be addressed, as was done in the case of the TRUST and PREPARED Codes. This approach identifies *only* real-world challenges, which serves as a crucial reality check, a key strength of the approach being that ethical requirements are grounded in actual risks and informed by diverse voices and experiences through extensive literature reviews, empirical work and consultations. Additionally, the broader the search for potential risks, the more likely it is that most will be identified. Hence, in our case, the great effort that was put into identifying the pandemic challenges for research ethics and research integrity over almost two years, across research disciplines, languages and cultures, contributed to the achievement of comprehensiveness. The risk-based approach has ensured that the PREPARED Code is both reflective of what happens in the real world and comprehensive.

3 Aligned with Core Moral Values

The risk-based approach offers a reliable way of telling us *what* needs to be addressed by a code of ethics, but it does not tell us anything about *how* these matters should be addressed. To ensure ethical decision-making, action-guiding codes of conduct must be grounded in a coherent moral framework. For the PREPARED Code, this framework is values-based, which involves the explicit adoption of specific moral values: fairness, respect, care, and honesty. These values guide decision-making and dispose the individual towards one course of action over another (Chatfield and Law 2024). While the choice of values may differ for other codes of ethics, the importance of alignment with core moral values should not be underestimated.

There are two main reasons why this is the case. First, a defining characteristic of values is their motivational power. This is especially true for values with explicit moral significance, which are often regarded as the most important (Schwartz 2012). Extensive empirical research on values has demonstrated that they play a crucial role in shaping behaviour, guiding decision-making and motivating individuals (Hitlin and Piliavin 2004; Illies and Reiter-Palmon 2004; Fritzsche and Oz 2007; Schwartz 2013).

Second, there is a significant body of research demonstrating that when people work in environments that are congruent with their core personal values, they assume greater personal responsibility, experience higher job satisfaction and enjoy improved wellbeing (Deci and Ryan 2000; Van Vianen 2000; Posner 2010; Schwartz and Sortheix 2018).

Thus, to motivate ethical action, it is important not only that codes are aligned with moral values, but also that these values resonate with the intended users of the code.

4 Transparent Development Process

Kaptein and Schwartz (2008) make the point, which we take further in Chap. 3, that knowing how a code was developed is a prerequisite to measuring its effectiveness. It must be clear who authored the code, and the rationale behind its creation must be transparent, because it is the behind-the-scenes *process* of code development that confers credibility (Messikomer and Cirka 2010). We therefore recommend that code authors document their development process carefully and make that documentation publicly available, just as we are doing through this book for the PREPARED Code.

5 Inclusive Development Process

Washington and Kuo (2020) emphasise that ethics codes often reflect the perspectives of those in power, which can have the effect of excluding perspectives from marginalised communities. They argue for incorporating diverse voices to ensure that ethics codes do not unintentionally prejudice groups in vulnerable situations. We believe that ethics codes should not be developed in isolation by an ad hoc group; a code is more likely to achieve credibility if the drafting process actively seeks and encourages broad participation (Messikomer and Cirka 2010). Engaging a diverse range of stakeholders in the development process helps create an ethics code that is comprehensive, equitably reflects diverse views and is culturally sensitive, ultimately securing its acceptance across different communities, research disciplines and geographic locations.

Inclusivity was central to the PREPARED Code's development, which incorporated diverse perspectives from across the globe, as shown in Fig. 2, reproduced here as Fig. 2.

Fig. 2. PREPARED Code authors: the geography

The process was further enriched by consultations with a wide range of stakeholders, including researchers, policymakers, research funders, publishers, NGOs and governance organisations. Notably, it also included input from communities worst afflicted during the pandemic (e.g. individuals on the poverty line and disabled people), ensuring that their perspectives were integrated into the ethics code.

Inclusivity also shaped every stage and aspect of evidence gathering, from working in multiple languages to engaging marginalised population groups through sensitive and appropriate methods. The PREPARED team actively sought dialogue with all groups that might be impacted by the code and encouraged discussion among them. By listening to the experiences and perspectives of a wide range of research stakeholders, the PREPARED team was able to co-create a code that will hopefully be widely acceptable to all involved in the research process.

6 Accessible

An accessible ethics code must be easy to understand and free of vague, complex or technical language. While we cannot say for sure that there is a direct correlation between the clarity of ethics codes and ethical behaviour, evidence exists that deficiencies in understanding contribute to research misconduct. For instance, in their qualitative interview study with scientists, Cairns et al. (2021) found that half of the participants referenced a lack of understanding of research ethics as a cause of unethical behaviour. The use of clear, unambiguous language in ethics guidance is therefore crucial.

The accessibility of any document can also be affected by its structure and length. For instance, the excessive length of an ethics document can discourage attempts to read it (Schwartz 2004). In Cameroon, for example, the important factors for research ethics procedures were identified as brevity, simplicity, clarity and user-friendliness.

> Whatever is brief and clear is better than what is not and saves time. What is simple and user-friendly is better than what is not even though the two have the same aims because it saves both time and mental energy. (Ouwe Missi Oukem-Boyer et al. 2016).

To enhance accessibility, the PREPARED team created a code that is concise, engaging and free of unnecessary jargon, thus ensuring clarity for researchers, funders, policymakers and public alike. This approach of making the PREPARED Code user-friendly, even for those without specialised knowledge of research ethics and research integrity, reflects a broader commitment to accessibility and transparency in research. Additionally, the Code was translated into twelve languages (Arabic, Chinese, Finnish, French, German, Greek, Italian, Korean, Lithuanian, Portuguese, Swahili and Spanish), thereby maximising its reach.

7 Implementable

An ethics code alone does not ensure ethical research (Nijhof et al. 2003). A real challenge for any new code is to raise awareness and demonstrate its practical applicability. The PREPARED team addressed this challenge by developing a range of resources designed to support the understanding and application of the PREPARED Code. These materials are not just informative but also encourage ethical reflection, prompting researchers to move beyond theoretical knowledge and actively engage with the real-world dilemmas they may encounter during crises.

The vital need for effective ethics training to complement any ethics code is broadly recognised (Schwartz 2004). Acknowledging that ethics training can often be dense and difficult to engage with (Miller-Dykeman n.d.), the authors of Chap. 6 share their insights on creating ethics training that tries to reflect the code's qualities of accessibility: concise, engaging, and free from jargon. They also offer practical strategies for developing such training and ensuring it reaches a global audience of researchers in a user-friendly and effective way. For instance:

- To ensure clarity and accessibility across diverse cultural, linguistic and geographic contexts, training clips were created to explain and contextualise each guidance article in the code.

- Recognising that a resilient research ecosystem requires the adaptation of existing processes, the team developed stakeholder-specific guidelines. These were designed to be aligned with the established procedures of research ethics committees, publishers, editors and research-performing institutions, facilitating seamless integration.
- To deepen engagement, the project prioritised interactive formats such as video clips, polls and discussions. Additionally, many training materials are available through the PREPARED mobile app, which offers global access and ease of use. The free PREPARED Case Study app presents research ethics training that is self-paced, engaging, interactive and conveniently packaged to enable smartphone access for both Android and iPhone users. The cases are mostly built on real-life examples, so as to be relatable to researchers, thereby increasing the likelihood of deep reflection, recall and application (Schroeder et al. 2025).

The authors encourage future ethics initiatives to go beyond simply drafting codes by offering practical tools that promote ethical reflection and accessibility.

8 Final Words

Ultimately, we hope that this code is never needed – that the devastation of the COVID-19 pandemic will not be repeated – but science warns us otherwise. The risk of a pandemic in the coming decades is ever-present and may be growing due to factors like urbanisation and climate change (Williams et al. 2023). According to Smith (2024), preparing for the next pandemic will require a blueprint to accelerate the organisation, coordination and conduct of critical research and development. This blueprint should be grounded in ethical commitments, standards and judgments that are capable of informing research priorities, collaboration and partnerships, and equitable data and benefit-sharing. It should also exemplify respect for all affected.

We are confident that the PREPARED code will be a valuable addition to this blueprint, through its strong ethical grounding, transparent and inclusive development process, and easy accessibility, aided by careful consideration of its future implementation through unique and innovative tools.

The final words in this book come from the lead author of the PREPARED Code, Prof. Doris Schroeder.

Standing on the Shoulders of Crowds – Ethically
Research ethics is a small, specialised field. Many people associate it with no more than a box-ticking exercise, a routine pit stop prior to setting off on the main track. It might surprise them to know that there are people whose research is actually *on* research ethics and research integrity. Perhaps they say to themselves: "How tedious! Don't they want to join the real race themselves?"

And it isn't only researchers who think this way, judging from the ethics offers in bookstores. The general public seem keen to read about a huge range of ethics topics, such as the ethics of hedonism, stoicism, climate responsibility or animal experimentation, or ethical issues in artificial intelligence. But will they want to read about research ethics or research integrity? No. At a stretch they might want to read about scandals. But who would want to read, in their leisure time, about data protection or about stopping the falsification of data?

That is what happens when a field becomes overly technical, when only technocratic elites can understand and contribute to discussions. In the context of politics, Michael Sandel (2020: 28) described it like this: "Our technocratic version of meritocracy severs the link between merit and moral judgment."

By making the TRUST and PREPARED Codes short, accessible, jargon-free, values-driven and co-created with groups in vulnerable situations, we want to open a door – a door from the pit stop onto the main track. In his Foreword, Michael Makanga has already opened it: "The world would be a better place if more human activities were governed by fairness, respect, care and honesty."

Professional ethics and how to conduct oneself, in our case as researchers, should not just be a technical study for a handful of people. It should be understandable to all. A *Nature* (2021) editorial described this aptly:

The metaphor "standing on the shoulders of giants" has been much overused by scientists ... Today, such "giants" are not only the investigators ... but also every other participant in the research process. The future lies in standing on the shoulders of crowds.

"Standing on the shoulders of crowds" means that mindsets and practices that are unfair, exploitative, and non-inclusive have to change. It means that the crowds must be equipped to understand and trust the ethical foundations of an activity. Fairness, respect, care and honesty could constitute such an ethical foundation, even for the main track itself, not just an ethics pit stop .

Doris Schroeder

Fig. 3. Fairness, respect, care and honesty

References

Cairns, A.C., Linville, C., Garcia, T., et al.: A phenomenographic study of scientists' beliefs about the causes of scientists' research misconduct. Res. Ethics **17**(4), 501–521 (2021). https://doi.org/10.1177/17470161211042658

Chatfield, K., Law, E.: 'I should do what?' Addressing research misconduct through values alignment. Res. Ethics **20**(2), 251–271 (2024). https://doi.org/10.1177/17470161231224481

Deci, E.L., Ryan, R.M.: The 'what' and 'why' of goal pursuits: human needs and the self-determination of behavior. Psychol. Inq. **11**(4), 227–268 (2000). https://doi.org/10.1207/S15327965PLI1104_01

Fritzsche, D., Oz, E.: Personal values' influence on the ethical dimension of decision making. J. Bus. Ethics **75**, 335–343 (2007). https://doi.org/10.1007/s10551-006-9256-5

Hitlin, S., Piliavin, J.A.: Values: reviving a dormant concept. Ann. Rev. Sociol. **30**(1), 359–393 (2004). https://doi.org/10.1146/annurev.soc.30.012703.110640

https://scienceandsociety.duke.edu/learn/ma/the-student-experience/profiles-graduates/engaging-young-scientists-in-research-ethics/. Accessed 6 Mar 2025

Illies, J.J., Reiter-Palmon, R.O.N.I.: The effects of type and level of personal involvement on information search and problem solving. J. Appl. Soc. Psychol. **34**(8), 1709–1729 (2004). https://doi.org/10.1111/j.1559-1816.2004.tb02794.x

Kaptein, M., Schwartz, M.S.: The effectiveness of business codes: a critical examination of existing studies and the development of an integrated research model. J. Bus. Ethics **77**, 111–127 (2008). https://doi.org/10.1007/s10551-006-9305-0

Messikomer, C.M., Cirka, C.C.: Constructing a code of ethics: an experiential case of a national professional organization. J. Bus. Ethics **95**, 55–71 (2010). https://doi.org/10.1007/s10551-009-0347-y

Miller-Dykeman, A.: Change is needed in research ethics training for young scientists. Alumni Reflections, Duke Science and Society (n.d.)

Nature: Research collaborations bring big rewards: the world needs more. Nature **594**(7863), 301–302 (2021). https://doi.org/10.1038/d41586-021-01581-z

Nijhof, A., Cludts, S., Fisscher, O., Laan, A.: Measuring the implementation of codes of conduct: an assessment method based on a process approach of the responsible organisation. J. Bus. Ethics **45**, 65–78 (2003). https://doi.org/10.1023/A:1024172412561

Ouwe Missi Oukem-Boyer, O., Munung, N.S., Tangwa, G.B.: Small is beautiful: demystifying and simplifying standard operating procedures: a model from the ethics review and consultancy committee of the Cameroon bioethics initiative. BMC Med. Ethics **17**(27), 1 (2016). https://doi.org/10.1186/s12910-016-0110-8

Posner, B.Z.: Another look at the impact of personal and organizational values congruency. J. Bus. Ethics **97**(4), 535–541 (2010). https://doi.org/10.1007/s10551-010-0530-1

Sandel, M.: The tyranny of Merit: What's Become of the Common Good? Farrar Straus and Giroux, Kindle edn. (2020)

Schroeder, D., Paspallis, N., Kasenides, N., et al.: The PREPARED case study app for deep reflection on research ethics challenges during crisis. Nat. Med. (2025, in press)

Schwartz, M.S.: Effective corporate codes of ethics: perceptions of code users. J. Bus. Ethics **55**, 321–341 (2004). https://doi.org/10.1007/s10551-004-2169-2

Schwartz, S.: Value priorities and behavior: applying a theory of integrated value systems. In: Seligman, C., Olson, J.M., Zanna, M.P. (eds.) The Psychology of Values, 1st edn. Lawrence Erlbaum Associates, Mahwah, pp. 1–24 (2013). https://doi.org/10.4324/9780203773857

Schwartz, S.H.: An overview of the Schwartz theory of basic values. Online Read. Psychol. Cult. **2**(1), 2307–0919 (2012). https://doi.org/10.9707/2307-0919.1116

Schwartz, S.H., Sortheix, F.: Values and subjective well-being. In: Diener, E., Oishi, S., Tay, L. (eds.) Handbook of Well-Being, pp 833–847. DEF Publishers, Salt Lake City (2018). https://static1.squarespace.com/static/65f0a38858b34640d8d1d19a/t/663ba4b32 297654ef513dcc0/1715184831926/Handbook-of-Well-Being.pdf. Accessed 6 Mar 2025

Smith, M.J.: Ethics of pandemic research. In: Sorenson, R.A. (ed.) Principles and Practice of Emergency Research Response, pp. 73–90. Springer, Cham (2024). https://doi.org/10.1007/978-3-031-48408-7_4

TRUST: The TRUST code: a global code of conduct for equitable research partnerships (2018). https://doi.org/10.48508/GCC/2018.05

Van Vianen, A.E.: Person-organization fit: the match between newcomers' and recruiters' preferences for organizational cultures. Pers. Psychol. **53**(1), 113–149 (2000). https://doi.org/10.1111/j.1744-6570.2000.tb00196.x

Washington, A.L., Kuo, R.: Whose side are ethics codes on? Power, responsibility and the social good. In: Proceedings of the 2020 Conference on Fairness, Accountability, and Transparency, pp. 230–240 (2020). https://doi.org/10.1145/3351095.3372844

Williams, B.A., Jones, C.H., Welch, V., et al.: Outlook of pandemic preparedness in a post-COVID-19 world. npj Vaccines **8**(178), 1–12 (2023). https://doi.org/10.1038/s41541-023-00773-0

Open Access This chapter is licensed under the terms of the Creative Commons Attribution 4.0 International License (http://creativecommons.org/licenses/by/4.0/), which permits use, sharing, adaptation, distribution and reproduction in any medium or format, as long as you give appropriate credit to the original author(s) and the source, provide a link to the Creative Commons license and indicate if changes were made.

The images or other third party material in this chapter are included in the chapter's Creative Commons license, unless indicated otherwise in a credit line to the material. If material is not included in the chapter's Creative Commons license and your intended use is not permitted by statutory regulation or exceeds the permitted use, you will need to obtain permission directly from the copyright holder.

Author Index

A
Andanda, Pamela 34

C
Chatfield, Kate 1, 16, 53, 94

D
Drummond, Orla 34

E
Evans, Natalie 53

F
Feinholz, Dafna 53

G
Gefenas, Eugenijus 16

I
Inguaggiato, Giulia 34

K
Kim, Ock-Joo 53
Kimani, Joshua 16
Kumar, Nandini 53

L
Law, Emma 16
Leisinger, Klaus 34
Lukaševičienė, Vilma 16, 34

M
Mlotshwa, Langelihle 34

N
Nyirenda, Thomas 76

O
Odhiambo, Joyce Adhiambo 16

P
Partington, Hazel 53
Paspallis, Nearchos 76
Perelló, Clàudia Pallisé 53

R
Robinson, David 8

S
Schroeder, Doris 8, 16, 53, 76
Seedall, Carly 76
Singh, Michelle 1, 76, 94

V
Videnoja, Kalle 16

Z
Zhu, Wei 53

The manufacturer's authorised representative in the EU is Springer Nature Customer Service Centre GmbH, Europaplatz 3, 69115 Heidelberg, Germany. If you have any concerns regarding our products, please contact ProductSafety@springernature.com

Printed and bound by CPI Group (UK) Ltd, Croydon, CR0 4YY

26/03/2026

02078969-0006